Informed and Healthy

Informed and Healthy
Theoretical and Applied Perspectives on the Value of Information to Health Care

Maria G. Nassali Musoke
Makerere University

AMSTERDAM • BOSTON • HEIDELBERG • LONDON
NEW YORK • OXFORD • PARIS • SAN DIEGO
SAN FRANCISCO • SINGAPORE • SYDNEY • TOKYO

Academic Press is an imprint of Elsevier

Academic Press is an imprint of Elsevier
125 London Wall, London EC2Y 5AS, UK
525 B Street, Suite 1800, San Diego, CA 92101-4495, USA
50 Hampshire Street, 5th Floor, Cambridge, MA 02139, USA
The Boulevard, Langford Lane, Kidlington, Oxford OX5 1GB, UK

Notices
Knowledge and best practice in this field are constantly changing. As new research and experience broaden our
understanding, changes in research methods, professional practices, or medical treatment may become necessary.

Practitioners and researchers must always rely on their own experience and knowledge in evaluating and using any
information, methods, compounds, or experiments described herein. In using such information or methods they should be
mindful of their own safety and the safety of others, including parties for whom they have a professional responsibility.

To the fullest extent of the law, neither the Publisher nor the authors, contributors, or editors, assume any liability for any
injury and/or damage to persons or property as a matter of products liability, negligence or otherwise, or from any use or
operation of any methods, products, instructions, or ideas contained in the material herein.

British Library Cataloguing-in-Publication Data
A catalogue record for this book is available from the British Library.

Library of Congress Cataloging-in-Publication Data
A catalog record for this book is available from the Library of Congress.

ISBN: 978-0-12-804290-8

For Information on all Academic Press publications
visit our website at http://www.elsevier.com/

Working together
to grow libraries in
developing countries

www.elsevier.com • www.bookaid.org

Publisher: Mica Haley
Acquisition Editor: Rafael Teixeira
Editorial Project Manager: Ana Claudia Garcia
Production Project Manager: Edward Taylor
Designer: Greg Harris

Typeset by MPS Limited, Chennai, India

Contents

Foreword by Ruhakana Rugunda...vii
Foreword by Ylann Schemm..ix
Dedication..x
Acknowledgments xi
Abbreviations...xiii

CHAPTER 1   Health Information in Uganda 1
   1.1 Introduction 1
   1.2 Information Concept 2
   1.3 Primary Health Care...4
   1.4 Information Technology in the Health Sector and Related Policies 8
   1.5 Health Information Challenges 11
   1.6 Organization of the Book...13
      1.6.1 Chapter 1: Health Information in Uganda 13
      1.6.2 Chapter 2: Access and Use of Information by Women and Health Workers.............13
      1.6.3 Chapter 3: The Value of Information and Effect on Health Outcomes......................14
      1.6.4 Chapter 4: Modeling Information Behavior 14
      1.6.5 Chapter 5: Implications for Theory, Practice and the Future......14
   References..14

CHAPTER 2   Access and Use of Information by Women and Health Workers 17
   2.1 Methodological Approach...17
   2.2 The Categories...21
      2.2.1 Information Sources 21
      2.2.2 Information Needs 21
      2.2.3 Constraints 22
      2.2.4 Moderators...22
      2.2.5 Value of Information 22
   2.3 Information Activities by Women 22
      2.3.1 Information Sources 22
      2.3.2 Information Needs 28
      2.3.3 Constraints 42
      2.3.4 Moderators...51
   2.4 Information Activities by Health Workers 59
      2.4.1 Information Sources 59
      2.4.2 Information Needs 70
      2.4.3 Constraints 78
      2.4.4 Moderators...84
   2.5 Discussion of Information Activities..90
      2.5.1 Information Sources 90
      2.5.2 Information Needs 94

2.5.3 Constraints 96
2.5.4 Moderators 100
2.6 Conclusion 102
References 102

CHAPTER 3   The Value of Information and Effect on Health Outcomes 105
3.1 Value of Information Core Category 105
3.2 Use of Information and Attribution of Value by Women 105
3.2.1 Value of Information 106
3.2.2 Actions 111
3.3 Use of Information and Attribution of Value by Health Workers 119
3.3.1 Value of Information 119
3.3.2 Actions 122
3.4 The Effect of Information on Health Care 126
3.4.1 Background 126
3.4.2 Evidence of the Effect of Access and Use of Information on Health Care 127
3.5 Discussion of the Value and Impact of Information 131
3.6 Conclusion 134
References 135

CHAPTER 4   Modeling Information Behavior 137
4.1 Introduction to the Model 137
4.2 The Core and Main Categories 137
4.3 Development of the Model: From Concrete to Abstract 139
4.3.1 Stage One 141
4.3.2 Stage Two 142
4.3.3 Stage Three 142
4.3.4 Stage Four 142
4.4 Discussion of the Model 142
4.5 Strengths and Limitations of the Model 150
4.5.1 Strengths 150
4.5.2 Limitations 151
References 152

CHAPTER 5   Implications for Theory, Practice and the Future 155
5.1 Introduction 155
5.2 Implications for Theory and Applicability of Findings 156
5.3 Implications for Health Informatics and Information Provision in General 158
5.3.1 Implications for Information Provision to Women 160
5.3.2 Implications for Information Provision to Health Workers 161
5.3.3 Implications for Health Information Professionals 162
5.4 Areas for Further Research 163
References 164

Index 167

Foreword

Librarians as agents of change: the transition to electronic health information

In the early 1990s, a small satellite orbited the Earth, pole to pole, quietly dropping off e-mail messages at least twice per day to computer and ham radio–based ground stations in Africa. This early, quite rudimentary Internet system was called HealthNet, and Albert Cook Library at Makerere University in Kampala, Uganda, was one of the sites where clinicians and researchers waited eagerly for the satellite to pass over.

Inside Albert Cook, library director Maria Musoke and her young assistant Sara Mbaga made sure the equipment on site was operating and secure, even building a cage around the ground station! They pulled down messages, all in ascii text, and saw that they were delivered (by hand) to the right people. They downloaded and printed the weekly HealthNet News, posted it on the library notice board, and circulated it to clinics, making the latest research available to clinicians and researchers at the university and beyond. These librarians were central to the success of HealthNet in Uganda.

I am happy to write the Foreword to Dr Musoke's book, as I saw this project start, grow and subsequently play an important role in arming both clinicians and researchers in Uganda with up-to-date information to improve their clinical and research performance. During my tenure as Minister of Transport and Communication, I was pleased to be part of this HealthNet team—as a government official who was trained as a pediatrician and had served as Minister of Health. The team also included well-known AIDS researcher Nelson Sewankambo (recently retired as the Principal of the College of Health Sciences at Makerere) and Charles Musisi (now a successful IT expert in the private sector).

I recount the story of those early days to illustrate that it was not enough to have Internet access, however nascent. The vast store of information that could eventually be tapped via the Internet called out to librarians as agents of change. These information activists reached out from the university to health workers—at the referral hospital, the rural clinic and a variety of sites in between.

The person who worked and continues to work tirelessly to move electronic information from the Internet and push it into the active service of improving health is Dr Maria Musoke, former Director of Albert Cook Library and University Librarian, now Makerere University Professor of Information Science and author of this book.

From the early days onward, Dr Musoke has been a leader in outreach, the reinvention and re-envisioning of how health information can be tailored for a variety of audiences. She thinks globally while acting locally in Uganda. In the field of health information access, she stands as both a midwife and a pioneer.

Her devotion to health information—its access and use—now moves across a very broad band, not only in the halls of academia but also in hospitals and in rural clinics in isolated or remote areas. Her mission has expanded beyond the technical intricacies of receiving information from a satellite and distributing it to clinicians and researchers to that of engaging librarians in the process of empowering people in communities—rural and urban—to find the information themselves.

She has traveled regularly with a team of doctors, nurses and midwives to train rural health workers, inform community members, and influence health outcomes. She has assisted librarians in repackaging information so that it supports field visits and helps local people with their health problems.

Her commitment is to translating knowledge into practice and to examining the value of information and its ability to affect the bottom line of better health. With this book, she offers a clear model for literacy and advocacy—how librarians in developing countries can reach out to rural health workers with great impact on health outcomes.

If you go to the Albert Cook Library today, you can still see those HealthNet antennae on the roof. They have long since given way to the inexorable movement of technology to very large satellites and undersea cable, but Dr Musoke insists that those antennae should remain on the library roof—as historic lightning rods that preserve the memory of those early days of Internet and health information access in Africa and continue to inspire us all.

Rt. Hon. Dr. Ruhakana Rugunda
Prime Minister of the Republic of Uganda

Foreword

Translating Knowledge into Health: Professor Musoke's Journey through Practice, Evidence and Policy in Uganda

Over the past decade, the Elsevier Foundation has supported capacity-building projects in the fields of science, technology and medicine through training, education, infrastructure and preservation of information. Professor Maria Musoke's project at Makerere University, **"Enhancing Access to Current Literature by Health Workers in Rural Uganda and Community Health Problem-Solving" was one of the most impactful and inspiring projects we ever had the honor of supporting. It embodied the underlying tenet of our program**: information access and services help professionals, researchers, teachers and students to be better educated and work more effectively, thereby improving population health and economic and social conditions. But how do you actually measure and quantify the impact that information has on health outcomes? This is notoriously difficult.

In her book, Professor Musoke has thoroughly explored this question using an in-depth qualitative methodology. For optimum clarity, she has also applied a gender lens, using her deep understanding of women's roles as health care information providers in rural Uganda. She has also used the 2001 intervention of free electronic access programs such as Research4Life's HINARI, offering large online libraries of top-quality current journals to explore changes in information use behavior and public health outcomes. The Elsevier Foundation has been proud to support Professor Musoke's work over the years to identify multi-angle approaches that may not be standard in the industrialized world but that are pioneering for health and information in a developing country environment. In her Elsevier Foundation project, Professor Musoke translated knowledge into practice for Uganda's rural health clinics. With her book, she has now effectively bridged practice to evidence for future researchers, practitioners and policy makers in this critical field. On behalf of the Elsevier Foundation: Bravo!

Ylann Schemm
Program Director, Elsevier Foundation

Dedication

To the dear ones who passed away after making various contributions to this process but did not live to see the end, namely:

- *My precious parents, Bridget and Hubert, who provided equitable educational opportunities to their daughters and sons.*
- *My loving spouse, John, who supported me throughout the process in various ways. Besides "mothering" the children, he read the different versions of this work and provided quantitative input to my qualitative research approach.*
- *My beloved sisters and brothers, who encouraged me in many ways. In particular, Winifred Jane Nantongo Jjuuko, the great teacher she was, for inspiring me to reach this far. She was and remains an excellent role model in my life. Secondly, Sarah Nalukwago Mukasa, who, during the beginning of this work, looked after my children very dearly.*
- *Professor John Mugerwa, Kulika—Uganda Scholarship committee member. He was one of the few administrators, at that time, who encouraged librarians to go for PhD studies.*

Acknowledgments

The funding from the Elsevier Foundation enabled me to respond to the many demands for this book and to realize my professional dream of writing a book on a topic very close to my heart. Special thanks to Ms Ylann Schemm, the Program Director, who coordinates the Innovative Libraries in Developing Countries program that sponsored this activity. I also do acknowledge, with many thanks, the initial support for my PhD work by the Kulika Charitable Trust and, later, the one-month Scholar's residency at Bellagio Center supported by the Rockefeller Foundation.

I thank Makerere University for the sabbatical. Many special thanks go to the Vice-Chancellor, Professor John Ddumba-Sentamu, and the Principal of CoCIS, Professor Constant Okello-Obura, for supporting my leave.

I greatly appreciate the University of Sheffield for hosting me during my sabbatical and conferring to me the title of "Visiting Academic" attached to the School of Health and Related Research (ScHARR) that provided a conducive environment for writing the book. Special thanks to Dr Andrew Booth, my immediate contact, for all the professional support and guidance and constructive comments that enabled me to complete the task on time.

Many thanks to my family for standing with and by me throughout this journey and for tolerating my frequent absence from home. Syl, Isaac and Miri kept reminding me that they are stakeholders in the final product. Professor Grace Ndeezi for being not only a family doctor but a mother to the children.

I am also grateful to Ms Julia Royall for being a true friend as far back as 1989, when she brought the good news to Sub-Saharan Africa of electronic mail that delivered current health literature to selected universities, including Makerere in Uganda. Since then, she continued supporting our efforts through the National Library of Medicine. We have witnessed the improvements in health information delivery, access and use through Julia's continued support even after her retirement.

Partnerships in Health Information (Phi) is greatly appreciated, particularly Ms Jean Shaw, who made comments on one of the difficult chapters at a time I desperately needed an unbiased eye. Secondly, Ms Shane Godbolt, who moved with me at the end of this long journey; it would not have been the same without the abundant support from Shane.

Several people provided various types of literature that I needed to write the book, and I am very grateful to them. There were times when I needed a piece of literature urgently and I contacted people who responded promptly, namely, Patrick, Brenda and Samuel in Uganda, and the ScHARR Resource center staff, particularly Magda, who made a difference in my urgent information needs. Furthermore, Brian Wamala "gathered" the various soft copies of documents that were scattered on different hard disks and put them in just two folders.

I do thank all the friends in and outside of Uganda who kept sending messages of encouragement to me, the prayers that I value very much, and other types of support. The "Maklib busmates" for keeping me updated. The Ugandans in the United Kingdom, particularly Ann and Simon, who made a difference in my stay in Sheffield. I thank you all very sincerely.

Finally, but by no means least, I thank the interviewees who participated in the study for their time and for providing the information that facilitated this work.

May God reward you all.

Abbreviations

CME	Continuing Medical Education
CPD	Continuing Professional Development
CUUL	Consortium of Uganda University Libraries
DDS	Document Delivery Services
DISH	Delivery of Improved Services for Health
DMO	District Medical Office, later became Directorate of District Health Services
DMUs	Dispensary Maternity Units
FP	Family planning
FPAU	Family Planning Association of Uganda
HINARI	Health InterNetwork Access to Research Initiative, part of Research for Life
HMIS	Health Management Information System
IEC	Information, Education and Communication
IFLA	International Federation of Library Associations
LCs	Local Councils = Local administrative authority/units from the village level to the District in Uganda
LIS	Library and Information Science
MakLib	Makerere University Library
MCH/FP	Maternal and Child Health/Family Planning under MoH
MoH	Ministry of Health
MUST	Mbarara University of Science and Technology
NGOs	Non-Governmental Organizations
NIDs	National Immunization Days
ORS	Oral Rehydration Salts
PHC	Primary Health Care
R4L	Research for Life (includes HINARI, AGORA, OARE and ARDI)
SDGs	Sustainable Development Goals
STIs	Sexually Transmitted Infections
TASO	The AIDS Support Organization
TBA(s)	Traditional Birth Attendant(s)
UCH	Uganda Chartered Healthnet
UHIN	Uganda Health Information Network
UMA	Uganda Medical Association
UNEPI	Uganda National Expanded Program for Immunization (under MoH)
UNFPA	United Nations Fund for Population Activities
UPE	Universal Primary Education
UPMA	Uganda Private Midwives' Association
WHO	World Health Organization

HEALTH INFORMATION IN UGANDA

1.1 INTRODUCTION

> the provision of health information to rural women is an asset to rural health care ... it helps us to improve personal, family and community health, and to reduce the incidences of disease
>
> **A woman from Lira district, Northern Uganda**

> we accessed information about the management of HIV/AIDS patients that recommended a combination of medication by adding Septrin that was not the case fifteen years ago, and now patients live longer and the quality of their life is much better than it was then
>
> **A clinical officer from a health centre in Masaka district, Central Uganda**

The quotes above, from a rural woman and a clinical officer working in rural Uganda, introduce the subject of the book by highlighting the value attributed to information by its users and the effect of information on health care.

Access to information is an essential component of development; it is a human right, and it does bring about sustained development and socioeconomic progress. However, over the years, Sub-Saharan Africa and other developing countries experienced some harsh economic policies that affected the social sector immensely. At the country level, the economic policies led to a reduction in public health spending. This had negative consequences on public health as a whole, but mainly its information sector. Consumers of health information faced various information accessibility challenges.

Setting up and implementing effective information services for rural people and the policies governing their use depend on ample knowledge of rural people's information environment and behavior. Hardly any empirical data exist on this topic in Uganda. Focusing research on rural health workers, who serve 80% of Uganda's population, is an important step in improving their information services and health care delivery in general. This would indirectly enhance the provision of information to the communities they serve and improve health outcomes in the end. At a local level, such research is important because the majority of Ugandans[1] do not see high-level health workers when they seek health care; it is provided within the family, community, or health units run by nurses and clinical officers.

---

[1]Uganda had a population of 34.857 million people in 2014, with slightly more females than males. Approximately 80% of the population live in rural areas. (*Source:* Uganda National Population and Housing Census Report, 2014).

Informed and Healthy. DOI: http://dx.doi.org/10.1016/B978-0-12-804290-8.00001-4

Although studies in the developed world tend to focus on information systems and retrieval (Lemberger and Morel, 2012; Dwivedi et al., 2012; Spink, 1999; Spink and Dee, 2007; Vakkari, 1999; Vakkari and Järvelin, 2005; Vakkari and Huuskonen, 2012), the interest of an information researcher in a rural African setting is inevitably information in everyday life. In a post civil war situation and after health epidemics in an African rural area, can the use of available information make a difference in the lives of rural people? Knowing that many diseases can be prevented by information/awareness of what to do or avoid or where to go, are people aware that information is important? Has the gradually increasing availability of information driven by technology been felt by the people? Is the provision of information by health workers to rural people or by information providers to health workers and rural people important? Is it important to health planners and policy makers, or does it become an issue only in times of epidemics? Does the notion "prevention is better than cure" still hold true in these situations? These thoughts or questions inspired the author to conduct a doctoral study titled "Health information access and use in rural Uganda" (Musoke, 2001). Having compiled several bibliographies, the author had identified a gap in existing studies on health and rural Uganda in particular and Sub-Saharan Africa in general. Furthermore, in 2014–15, the author conducted a follow-up study to update the previous findings where necessary.

This book draws extensively from the findings of the aforementioned studies, whose main aim was to investigate the accessibility and use of health information by women and health workers, who are at the lowest level of Primary Health Care (PHC) service delivery in Uganda.

1.2 INFORMATION CONCEPT

Information researchers have generally expressed the difficulties of defining or conceptualizing information (Roberts, 1976; Dervin, 1977; Belkin, 1978; Wilson, 1981; Browne, 1997). Information may be taken as the raw material from which knowledge is derived. To be utilized, information, like other resources, has to be tapped. In practice, however, this has been difficult to achieve in many African countries because of various factors, such as financial, sociocultural and lack of a supportive policy (Musoke, 2007a; Gadau & Lwoga, 2013). Even in the developed world, cognitive barriers have been reported to affect information access and use (Savolainen, 2015).

In Uganda where the research was conducted, however, the word "information" does not exist in various local languages. Consequently, the words "knowledge," "happenings" and/or "news" are used to refer to information. To put information in the right context, one has to phrase a sentence that would make the word information different from the ordinary usage of the words knowledge, news, or happenings. This was reflected in the way some questions were worded or phrased in the interviews conducted in the vernacular with women. Other authors have shared similar experiences; for example, Namyalo (2002) highlighted the challenge of translating health messages in a multilingual and multicultural Uganda where some words are difficult to correctly translate from English because they do not exist in the vernacular, and yet Uganda has no national language.

For a long time, many authors have recognized the fact that the term "information" is used in many different contexts; therefore, there is a need for an agreed-on concept of information. Furthermore, a suitable concept of information is necessary for both theoretical and practical development in information science. Authors such as Roberts (1976, 1982) and Belkin (1978) developed concepts that they considered suitable or proposed that existing information concepts from other fields could be applied to the context of information science. A list of eight requirements that are definitional, behavioral and methodological was elaborated by Belkin (1978: 59–62). The variety of frameworks used led to information for information science being variously considered, for example: a fundamental category such as matter; a property of matter; the probability of the occurrence of an event; a raw material from which knowledge is derived; reduction in the degree of uncertainty in a state of knowledge or similar construct; an event that takes place when a recipient encounters a text; data of value in decision-making; publicly communicated scientific information and simply information as message.

Dervin (1977) conceptualized information in three ways: objective information that describes external reality; subjective information that describes internal reality and information as the behaviors enacted by the individual in the process of understanding reality. Dervin therefore brought on board the objective and subjective aspects of information as well as the behavior.

Information has also been regarded as a commodity because it can be bought and sold (Browne, 1997). Although it is true that various types of information, such as printed, audio-visual and/or electronic sources, are bought and sold, the commoditization of information, in that sense, is something this book may not emphasize because there are several other intervening factors (Apalayine and Ehikhamenor, 1996; Musoke, 2001).

Generally agreeing with Dervin's (1977) conceptualization and Browne's (1997) definition, Braman (1989) provided a hierarchical set of definitions comprising four groups: information as a resource, information as a commodity, information as a perception and information as a constitutive force in society. The first two definitions are generally objective. The third definition, perception, refers to information as intangible and subjective. The last one combines the first three and has some behavioral perspectives.

Since then, some of the aforementioned concepts have become commonly used in information work. For example, Wilson (1999) reported that the purpose of his study was to explore information behavior based on the "problem-solving process and upon the concept of uncertainty reduction." Ross's (1999) definition of information was also close to that: "We adopt a definition of information as something that fills in a gap in understanding or makes a difference to an individual's cognitive structure or helps people with their lives" (Ross, 1999: 343).

Other concepts, however, have been dropped. Several authors (Belkin, 1978; Wilson, 1981) reported that the most commonly proposed information concept for information science, at that time, was Shannon's "mathematical theory of communication": source, transmitter, receiver and destination, with message and noise. This was originally proposed for telecommunication. However, the concepts were found not to fulfill the requirements for information science concepts: "Shannon's information measure refers not to the message itself in terms of its contents, but rather to the probabilities assigned by the potential recipient to the set of all possible messages ... it is an extremely limited view of information, and one which might be difficult to apply to the context of information science, where information is traditionally associated with the meaning of a message, rather than the probability (or improbability) of its receipt. In particular, this concept of information fails to meet any but the last two of the requirements of an information concept for information science, for it explicitly aims not to consider meaningful, social communication, or the problems raised by the requirements which refer to the effect of information and the relationships between information and state of knowledge" Belkin (1978: 66).

Shannon's model or concepts could not apply to a study of information access and use. The situations in which information is accessed and used are generally social; hence, the study had to focus particular attention on the social context in which information processes take place. Furthermore, information use usually leads to changes in the user's state of knowledge, which involves interpretation and putting meaning to information, and yet these are issues Shannon's model tended to ignore.

Hence, the author conceptualized information in the context of user studies research following Wilson (1981), who pointed out that the problem with the information concept may not lie so much in the lack of a single definition as it does with a failure to use a definition appropriate to the level and purpose of the study. The term "health information" is used to refer to information on or about health and is not limited to epidemiological or statistical information.

In this book, the concept information is, therefore, used in a broad sense to include the objective and subjective aspects of information, for example, information as a physical entity (eg, books and posters), as facts, advice and opinions (printed, electronic, oral, audio or visual). Dervin's (1977) conceptualization of information is used in the book, namely, objective and subjective information as well as information behavior of the individuals.

1.3 PRIMARY HEALTH CARE

Since the 1978 international conference in Alma-Ata (USSR), many developing countries, including all Sub-Saharan African countries, adopted a PHC strategy after endorsing the Alma-Ata Declaration. The declaration defined PHC as follows: "Primary health care is essential health care based on practical, scientifically sound and socially acceptable methods and technology made universally accessible to individuals and families in the community through their full participation and at a cost that the community and country can afford to maintain at every stage of their development in the spirit of self reliance and self determination. It forms an integral part both of the country's health system, of which it is the central function and main focus, and of the overall social and economic development of the community. It is the first level of contact of individuals, the family and community with the national health system bringing health care as close as possible to where people live and work, and constitutes the first element of a continuing health care process" (WHO, 1978: section VI).

The basic features of PHC, therefore, are accessibility to individuals, availability to potential users and consistency with the principles of preventive, curative and rehabilitative health service delivery. The main task for each country participating in PHC is to develop a health system that is country-, situation- and problem-specific and that would, in the most efficient and effective way, try to solve priority health problems and contribute to improving the health status. In Uganda, this led to, among other things, decentralization of health services to district levels and integration and planned improvement of services up to rural areas, which included the setting up of Village Health Teams[2] at the lowest level in the communities.

---

[2]The Village Health Teams provide nonprescriptive items such as deworming medicines, condoms and family planning medications to those who have been using them.

Decentralization in Uganda, therefore, meant that the everyday running of the health services was no longer the responsibility of the Ministry of Health (MoH) because it had been devolved to the districts (Government of Uganda, 1997). The decentralized responsibility for running health care and for health planning was devolved to the district health teams, led by the Director of District Health Services. In the PHC setup, therefore, the district is the critical level or unit to perform the microactivities. This left the MoH with the following functions:

  i. Formulation of health policy and national planning for health;
 ii. Setting of standards and targets and monitoring them;
iii. Human resource development by organizing education and training;
 iv. Overseeing operational research into appropriate health care;
  v. Coordination of international agencies and
 vi. Overall national supervision.

Globally, PHC has 10 essential elements or activities within which its programs are recommended to be implemented (WHO, 1978). However, following Uganda's recognition of its national needs, the last element on the list here was added; this raised the total to 11. They are:

   i. Education of individuals and communities regarding prevailing health problems and methods of preventing them;
  ii. Promotion of food supply and proper nutrition for all community members;
 iii. Adequate supply of safe water and basic sanitation;
  iv. Maternal and child health care, including family planning;
   v. Immunization against major infectious diseases;
  vi. Prevention and control of locally endemic diseases;
 vii. Appropriate treatment of common disease and injuries;
viii. Provision of essential drugs;
  ix. Provision of mental health services;
   x. Provision of dental health services and
  xi. Provision of rehabilitative health services.

These services can be provided both in the community and in health units (see PHC Fig. 1.1).

At a national level, women have been recognized both as providers and promoters of health through their reproductive and productive roles and related activities. As health providers in the Ugandan family, the care or treatment provided by women at home is very important because it is the first response to an illness episode. Hence, its success or failure can affect the course and eventual outcome of the illness. Given the problems of accessibility to health units/health workers, care provided at home has become central to many people's lives. It is not surprising that a holistic PHC setup starts with "care by the family." In the Ugandan family, it is the women who generally provide that care.

The services or activities of PHC are provided within a framework of five principles:

i. *Equitable distribution and political commitment*: these require a more equitable distribution of health resources, both within districts and the country, and between them. In practical terms, they require special attention to the needs of the vulnerable or socially disadvantaged groups. Equally required is the political will for supportive policy, resource allocation, community mobilization and general support of PHC.

ii. *Community participation*: the active involvement of local communities is required at different stages of PHC for firm integration of the program in the community, socially and developmentally and with the support of the health services.

iii. *Focus on prevention*: this recommends a holistic approach to health that makes prevention equally important as cure in a continuum of care that extends throughout the life span shift with emphasis on health promotion and disease prevention.

iv. *Multisectoral approach*: health workers, in addition to their health care roles, must relate to both communities and other sectors as a team for improving all areas that influence health, even when they are beyond the traditional scope of the health services.

v. *Appropriate technology*: for fostering community ownership of the innovation, the required technology for the management of PHC needs to be in place.

The Alma-Ata Declaration also recognized that the existing medical training in both undergraduate and postgraduate courses did not adequately provide the necessary skills and attitudes required for PHC. This was mainly because medical training was primarily hospital-based. Yet many of the problems faced in the PHC situation were hardly seen in the teaching hospitals. As a result, universities were asked to develop postgraduate courses in the areas of PHC, and to develop curriculum that would begin to expose the students to the PHC situation in their undergraduate years. Hence, in 1989, Makerere University, the leading university in Uganda, introduced a new three-year postgraduate program, leading to the award of the degree of Master of Medicine in Community Practice. Later in 2003, it also introduced a community-based education and service program (COBES) for undergraduate medical students to prepare them for work in rural settings, while building the capacity of rural communities to promote health and prevent diseases through self-reliant approaches. Similarly, an undergraduate MBChB degree with a focus on community practice was started at Mbarara University of Science and Technology (MUST) in Western Uganda.

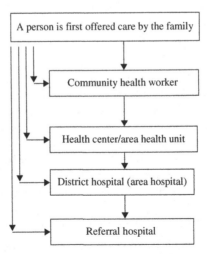

FIGURE 1.1

Primary Health Care service delivery components.

Musoke, M.G.N. (Ed.), 1997. The Uganda Health Information Digest 1 (1). ISSN: 1028–5105.

In 2008, during the 30th anniversary of the Alma-Ata International Conference that put health equity on the international political agenda for the first time, the World Health Organization (WHO) renewed its campaign to focus on PHC more "now than ever before." This was because "globalization is putting the social cohesion of many countries under stress, and health systems are clearly not performing as well as they could and should. People are increasingly impatient with the inability of health services to deliver. Few would disagree that health systems need to respond better—and faster—to the challenges of a changing world. PHC can do that" (WHO, 2008). The report went on to point out that "High maternal, infant, and under-five mortality often indicates lack of access to basic services such as clean water and sanitation, immunizations, and proper nutrition." WHO (2008) also noted that conditions of "inequitable access, impoverishing costs, and erosion of trust in health care constitute a threat to social stability." To steer health systems toward better performance, there was a need to return to PHC. When countries at the same level of economic development were compared, those where health care was organized around the tenets of PHC produced a higher level of health for the same investment. Such lessons were very important given the global financial crisis.

As initially articulated, PHC revolutionized the way health was interpreted and radically altered prevailing models for organizing and delivering care. It represented a deliberate effort to counter trends responsible for the "gross inequality in the health status of populations in both rich and poor countries. Inequalities in health outcomes and access to care are much greater today than they were in 1978." WHO (2008) estimated that better use of existing preventive measures could reduce the global burden of disease by as much as 70%.

In rural parts of Uganda and some other parts of the developing world, there was a tendency to fragment health care into discrete initiatives focused on individual diseases or projects, with little attention to coherence and little investment in basic infrastructures, services and staff. Furthermore, the existing health care was failing to respond to increasing social expectations for a health care service that was people-centered, fair, affordable and efficient. Implementation of a PHC setup was expected to bring balance back to health care and to put families and communities at the hub of the health system that they would create, own and sustain.

Furthermore, PHC would also offer the best way of coping with the ills of life in the 21st century, for example, the globalization of unhealthy lifestyles and the rapid unplanned urbanization. These trends contribute to an increase in chronic diseases, such as heart disease, stroke, cancer, diabetes and asthma, that create new demands for and resultant costs of long-term care and strong community support. A multisectoral approach is central to prevention because the main risk factors for such diseases lie outside the health sector. WHO pointed out that health systems will not naturally gravitate toward greater fairness and efficiency. Deliberate policy decisions are needed, such as: universal coverage reforms to reduce exclusion and social disparities in health; service (including information) delivery reforms aimed to organize health services around people's needs and expectations; public policy reforms to integrate health into all sectors; leadership reforms to pursue collaborative models of policy dialog; and increasing stakeholder participation.

The ultimate goal of PHC is better health for all. WHO encouraged countries to revert to the basics and use the 30 years of well-monitored PHC experience that had shown what works in rich and poor countries alike.

This book, which focuses on information processes at the lower levels of PHC service delivery in a Sub-Saharan African environment that has not been previously catered to by researchers or authors, is therefore timely.

1.4 INFORMATION TECHNOLOGY IN THE HEALTH SECTOR AND RELATED POLICIES

In the current era, technology has taken the fore front as the major enabler of information access by the information user. Several scholars have demonstrated that it is now almost impossible to study information without focusing on information and communication technologies (ICTs) because they appear in the "product" and "process" life history of information (Wilson, quoted by Adem (1997)). Other authors have similarly observed that it is difficult to disentangle ICTs from information. Mutch, for example, noted that "Indeed, many treatments which promise to be about information are on a close reading about technology exploitation. No serious treatment of information or IT can fail to take account of the inter-relationship and interdependence between the two, but there is a need to treat them as analytically distinct categories in order to examine such relationships" (Mutch, 1999: 535).

That recommendation was taken into consideration in this book. Among other things, ICTs and information share in common "the information user," who is the focus of the book.

The Uganda MoH was equipped with IT facilities, an automated health information system, and radio call equipment mainly for communication with upcountry stations. Furthermore, to address the challenge of limited flow of information from the central level to the districts and the lower levels in the community, the Uganda MoH made use of increasing mobile telephone facilities and put in place various communication channels to the districts and lower levels. This greatly changed the communication gap challenge that existed in the past, and agreed with what Kenney (1995: 36) had observed earlier: "it is particularly in the poor and isolated areas that electronic information delivery mechanisms can compensate, to a significant degree, for the isolated condition of millions of people.... Information delivered via telecommunications systems can be, and is being, used in various developing areas of the world for medical consultation to provide diagnostic and treatment support to relatively unsophisticated rural health workers or to the isolated patients themselves in particular cases."

The MoH received an Africa eHealth award in May 2015 for the communication system it put in place, which was titled "mTrac: Real-time monitoring and evaluation of disease surveillance, ACT Drug stock and health service delivery."[3] In the health sector, technology offers the opportunity for more transparency and accountability in service delivery as well as evidence-based practice and error reduction, diagnostic accuracy and treatment. It also facilitates client empowerment, enabling better self-care and health decision-making. It could also be used to shift tasks down the skills ladder, thereby addressing skills shortages. It also raises cost efficiency by streamlining processes, reducing waiting times and improving accuracy of data.

Within the MoH, several departments were responsible for health information collection, management and dissemination. The Health Planning Unit, for example, collected data and information from hospitals and other health units in the country, and from research; it also compiled statistics and ran the Health Information Management System (HMIS) of the Ministry. The MoH recognized, in its policy document, that it was essential for efficient planning and management to have accurate information on health and on the use of financial and other resources.

---

[3]https://www.mustaphamugisa.com/health/276-uganda-s-ministry-of-health-wins-africa-ehealth-award. Retrieved May 4, 2015.

In addition, the Uganda national health policy included information, education and communication as one of its goals. Furthermore, the national population policy stated that "Advocacy and information, education and communication, along with reproductive health, have emerged as key elements in the implementation of national population programmes." In addition, the Uganda Ministry of Information and Broadcasting had the mandate to inform, educate and entertain; therefore, it worked with the MoH and the media to effect its work in health.

Furthermore, a Uganda Communications Act became effective in 1997. Its objective was to develop a modern communications sector and infrastructure by, among other things:

i. Enhancing national coverage of communications services and products, with emphasis on provision of communication services;
ii. Expanding the existing variety of communications services available in Uganda to include modern and innovative postal and telecommunications services;
iii. Encouraging the participation of private investors in the development of the sector;
iv. Establishing and administering a fund for rural communication development.

The national communication network was generally strengthened through the liberalization of the sector. The liberalized communication policy made it easy for people to access information with hardly any restrictions. It is expected that in the future, rural communities would become better connected to reduce the gap between rural and urban connectivity.

The Uganda Communications Commission indicated that there were 5,700,000 Internet users in June 2012 in Uganda, 6,800,000 in June 2013 and 8,531,081 in June 2014. The majority of the users were, however, in urban areas and mainly in universities and other tertiary institutions, research organizations, Government departments and related institutions. The rural areas contributed a small percentage of the total Internet users.

Similar issues were raised in the literature about the Internet and connectivity in Sub-Saharan Africa. Lowan et al. (1998) and later Ajuh (2008) discussed the potential as well as the challenges. They pointed out that the Internet would bridge the information gap that was so apparent in developing countries' health and other institutions. However, access factors, the information on the Internet being predominantly North-oriented, and IT literacy were still major problems even though there was full Internet connectivity in all the African capital cities.

These observations were confirmed by the World Information Society report (2014) that pointed out that the digital divides still existed and some people were still excluded from access to communication networks. With a largely rural population, Uganda's IT infrastructure required urgent extension to rural areas. By 2015, and using the communications policy, the Government of Uganda had started implementing plans to improve services to rural areas.

It was gratifying to note some efforts in the right direction; for example, the "Technology, People and Process" (TPP) program of the Uganda Ministry of ICT had raised the IT literacy of Ugandans such that most of the students joining universities and other tertiary institutions in 2014—15 were more computer literate than those who joined ten (or more) years ago. This was further complimented by the increasing use of smartphones to access online information resources. Consequently, there was increased demand for IT-driven information services unlike in the past when students admitted to universities were mostly not computer literate. When such students graduate and run health services, for example, they continue making use of current online information resources to guide their clinical and related decisions that improve health care.

One of the institutions that provided information support to practitioners in Uganda was Makerere University Library (Maklib), which had ten branch libraries, including one in the College of Health Sciences known as Albert Cook Library. Maklib served as the Uganda national electronic resources coordinator in addition to its role as an academic library. Furthermore, Maklib continued to work closely with the Consortium of Uganda University Libraries (CUUL) to provide country access to a broad range of electronic information resources via institutional subscriptions and through collaboration and/or global initiatives that provide free access to resources by universities, research institutions and other not-for-profit institutions such as health units in Uganda. The resources accessed include Research for Life (R4L), initiated by Kofi Annan and launched in July 2001 to benefit low-income countries such as Uganda. R4L resources focus on four major subjects, namely health (Health InterNetwork Access to Research Initiative (HINARI)), agriculture (Access to Global online Research in Agriculture (AGORA)), environment (Online Access to Research in the Environment (OARE)) and technical (Access to Research for Development and Innovation (ARDI)). By 2014, there were sixty-four institutions registered for R4L resources in Uganda.

Before the R4L initiative, there was the Uganda Health Information Network (UHIN) implemented by the Uganda Chartered Healthnet (UCH) at the Makerere University College of Health Sciences (Mak-CHS) in Uganda. UCH was formerly Healthnet Uganda, which was responsible for the implementation of the first electronic health information service with the Albert Cook Library in the late 1980s, as outlined by Hon Rugunda in the Foreword. The experience of hosting and implementing that exciting information service, from the librarian's point of view, was shared at various fora (Musoke, 2007c, 2009). In addition to serving the Mak-CHS students, researchers and academics, the other specific objective of UHIN was to support health workers in selected districts to improve the quality of health care by providing them with relevant health information, in print and electronic formats, on prevention, diagnosis, treatment and general patient care related to major health problems of the districts. UCH continued to deliver relevant and timely continuing professional development materials on topical issues selected by the district health service teams in consultation with the MoH.

Although there were a number of projects related to evidence-based medicine, telemedicine, telehealth and eHealth in general in Sub-Saharan Africa, there was scanty literature on these topics. It is expected that some literature on the topics from the region will soon be available, given the ongoing research and the drive by the International Federation of Library Associations (IFLA) to document the contribution of libraries and librarians to the Sustainable Development Goals.

However, there were some IT-related studies on health from the Sub-Saharan African region and several studies on health information systems (Baldeh, 1997; Gladwin, 1999) conducted in the Gambia and Kenya, the Gambia and Uganda and Ghana, respectively. There were also the more general IT studies with sections on health, such as that of Adem (1997), the use of medical resources websites by university medical students and staff (Rhine and Kanyengo, 2000), the use of computers by medical students (Ajuwon, 2004), usage of electronic resources and information seeking behavior of medical library users (Musoke and Kinengyere, 2008; Chande-Mallya, 2014), respectively. Such studies informed the author in several ways. The studies on health information systems focused on district hospitals or district health staff/teams. This is understandable because the systems that were being studied were in the district hospitals and this may also explain why Forster (1990) used a case study method. Gladwin (1999) also used a case study method but focused on district health teams. Furthermore, Baldeh (1997) undertook action research to "demonstrate the use of an integrated health information system at the district level." The general study of IT was also urban-based because, by the time of the study, this is where the technologies were mainly used in Sub-Saharan Africa. The rest of the studies by Rhine and Kanyengo (2000), Ajuwon (2004), and, most recently, Gadau and Lwoga (2013) and Chande-Mallya (2014) focused on university medical students, physicians and other staff in urban-based teaching hospitals or universities/medical libraries. Hence, apart from simple technologies such as radio and mobile phones, the more advanced ITs such as the Internet were reported to be concentrated in African capital cities. Upcountry towns and rural areas seemed to be faced with challenges of poor connectivity. This book shares its findings of what was happening in rural Uganda.

Concluding this section, it was noted that the IT-driven information sector in Uganda was slowly but steadily improving even though there was a marked imbalance between the rural and urban areas. In the future, the general improvement in the Ugandan economy, the Poverty Eradication Action Plan (PEAP) that prioritized investments in social services including health, and the Health Sector—wide Approach that aligns available resources to specific priorities in the Health Strategic Plan are expected to lead to further improvements in the state of health informatics and general health information provisions in the country.

1.5 HEALTH INFORMATION CHALLENGES

Over the years, information-related challenges in the health sector have raised the concerns of information professionals, health workers and social scientists. Some suggested that to implement the planned health strategies, there would need to be an improvement in not only health information services but also the understanding of why and how to use the information. Several authors have reported that information was available but not accessible to many health professionals and librarians in Sub-Saharan Africa; however, for some, information was neither available nor accessible (Lowan et al., 1998; Musoke, 2007a). Although there was a need to produce more relevant information in Sub-Saharan Africa, the greatest challenge was to ensure that what was available so far could be accessed and used. By 2015, the situation had gradually changed with increased access to the Internet, particularly in urban areas of Uganda and the slowly improving information technology infrastructure, although the cost of bandwidth was still a major limiting factor.

A large body of literature on user studies has existed since the late 1940s, but the progress toward some theoretical understanding of key concepts such as information use has been slow, mainly because of inadequate methodology. By 2015, when this book was prepared, literature on information needs and information seeking behavior within the field of library and information science had extended into several hundred reports and journal articles. Generally, there was more and better conceptualization in the field.

It was noted that although user studies had grown tremendously in quantity and had also improved in quality within the past three decades, a review of literature highlighted gaps in the issue of value attributed to information by users and its effect on health care. There were also gaps in the understanding of health information access and use in a Sub-Saharan African setting and the lower levels of PHC. Furthermore, there was scanty literature on user studies in Uganda because rural women and lower-level health workers' information activities had not attracted information researchers. This was also true in some other parts. For example, Goel et al. (2015) reported that their study was the first to examine health information needs of community health workers in northern India. The assessment of the effect of information on health outcomes is an issue of concern to many development agencies and planners. The book highlights the many positive effects of increased access and use of information on health care in a low-income country. There were hardly any research-based and comprehensive works in the literature on this topic in Sub-Saharan Africa. Even in other parts, most literature on value of information focuses on the tangible economic aspects rather than the outcomes, such as the ones presented in the book. This book has bridged a gap by disseminating research findings on a topical issue in a Sub-Saharan African setting, which can be applied to other developing countries as well as the underserved communities in developed countries.

Although some previous studies on topics related to this book had a number of shortcomings, the majority were useful in indicating gaps and issues that needed further or fresh research. Some contributed topics or themes that were included in the study, namely, information use, needs and sources. Furthermore, access factors were included after Barton and Wamai (1994) raised concerns about the communication gap between the center and the communities in Uganda. Later, Kapiriri and Bondy (2006) also indicated that "the frequency of use of the different sources of information necessitates research to understand the barriers and careful planning of health information delivery to ensure equitable access." Then, Paek et al. (2008) indicated that research should identify factors that are associated with the roles of social trust and social network on family planning and other health behavior. Outside Uganda, Justice (1984) indicated that there is a need to understand, from a qualitative perspective, the influence of sociocultural practices on health behavior and on the delivery of PHC at a local level. Similarly, Apalayine and Ehikhamenor (1996) and Momodu (2002) highlighted beliefs and taboos and other social-cultural factors as being crucial for health information delivery in the West African rural communities studied. Hence, an open question about factors affecting access to information was included in the study.

The author took into account issues arising from the literature and addressed them by broadening the focus to include identified gaps or topics in the study using a qualitative method of data collection and an interpretive approach to data analysis.

A fundamental requirement for information acquisition is that some source of information should be accessible. Hence, this book focuses on sources of information, needs for information and access factors. Furthermore, the book focuses on what happens when one accesses information—is information used or not and why? It has been further noted that when people access and use information, there is some intervening user behavior; hence, the book inevitably focuses on information behavior.

As elaborated in Section 1.2, the term "information" is used to include the subjective and objective aspects of information as well as the behaviors associated with information acquisition and use based on the conceptualization of information by Dervin, 1977.

An investigation of access to and use of health information in rural Uganda is an applied piece of research, the purpose of which is to understand the issues being studied to generate actual and/or potential solutions to human and societal concerns or challenges in health. The findings have contributed knowledge that will enable planners, policy makers, development agencies, health-related organizations, information workers, health workers, academics and researchers in libraries, information, informatics, health communication, information systems and public health, as well as society itself, to understand the nature and sources of a problem so that human beings can more effectively access and use health information for the betterment of their health. While in basic research[4] the source of questions are the traditions within a scholarly discipline, in applied research the source of questions are the problems and concerns experienced by people.

The study on which the book is based was, therefore, holistic and user-centered. A holistic inductive paradigm used in the study provided a methodological foundation for obtaining the kind of data concerning women and health workers' information access and use patterns that have been used as the basis for deriving a new model of information access and use in rural Uganda (Musoke, 2001, 2007b). Having been inductively derived from data, such a model offers insight, enhances understanding and provides a meaningful guide to action and future planning to enhance health communication, informatics and health education and promotion in general.

The general information behavior models such as that of Wilson (1997) have an inclination toward information seeking behavior. Later, Wilson (1999) confirmed that there were few models of information behavior in general. Literature from Sub-Saharan Africa hardly provided any model of health information access and use. Generally, there was a relative scarcity of theoretical and methodological paradigms.

The new model of information behavior presented in this book is the main contribution to knowledge from the research. The model advances the understanding of the role of information to health care and the information processes involved. The main concern of the book, therefore, is not to quantify data, but rather to advance the understanding of issues surrounding access and use of health information in a Sub-Saharan African setting and how women and health workers perceive and interpret these issues. The implications of this work for information provision and theory are presented in the last chapter of the book. The rest of the chapters are outlined in the next section.

1.6 ORGANIZATION OF THE BOOK

The book has five chapters as outlined in this sub section. All the works referred to are listed at the end of each chapter in the book.

1.6.1 CHAPTER 1: HEALTH INFORMATION IN UGANDA

It introduces the book by presenting some background information that puts the focus of the book in context.

1.6.2 CHAPTER 2: ACCESS AND USE OF INFORMATION BY WOMEN AND HEALTH WORKERS

The chapter discusses the holistic inductive paradigm applied in the research and the Grounded theory approach used in the data analysis. It then provides the research evidence on which the information model was built by presenting the findings from women and health workers.

---

[4]The "purpose of basic research is knowledge for the sake of knowledge." The difference is that applied researchers are trying to understand how to deal with a problem while basic researchers are trying to understand and explain the basic nature of some phenomenon (Patton, 2002).

Chapter Two finally draws together the main findings and discusses them with reference to relevant literature, making it the longest chapter in the book.

1.6.3 CHAPTER 3: THE VALUE OF INFORMATION AND EFFECT ON HEALTH OUTCOMES

The chapter highlights the value attributed to information by women and health workers and its effect on health care. The emergent but core category, namely, value of information, is one major focus of this book, and it is presented and discussed in this chapter. After providing the research evidence in Chapters Two and Three, the book then proceeds to present the second major contribution to knowledge, the information model, in the next chapter.

1.6.4 CHAPTER 4: MODELING INFORMATION BEHAVIOR

The chapter presents the information access and use model. The model was initially made up of three emergent and two root categories that, after further analysis and abstraction, resulted in one core and two main categories. The chapter presents the stages of developing the model from concrete to abstract, which led to what was finally conceptualized as an Interaction-Value model. Chapter Four then concludes by highlighting the strengths of the model and reflecting on the limitations of using a holistic inductive approach and a Grounded theory in a Ugandan setting. The theoretical perspectives that pave the way to the future are presented in the last chapter.

1.6.5 CHAPTER 5: IMPLICATIONS FOR THEORY, PRACTICE AND THE FUTURE

The author concludes the book, in this chapter, by highlighting the implications of the Interaction-Value model for theory and/or information behavior models and discussing whether the findings can be applied, replicated or transferred elsewhere. Concluding comments are presented regarding health informatics and the provision of information in general. Finally, areas for further research are outlined.

REFERENCES

Ajuh, J., 2008. The persistence of rural poverty in Cameroon: urban bias in information and communication technology (ICT) development. In: ICT4Africa 2008 Conference Proceedings (www.ictforafrica.org). Visited March 14, 2010.

Ajuwon, G., 2004. Use of computers and the Internet in a Nigerian teaching hospital. J. Hosp. Librariansh. 4 (4), 73–88.

Adem, L., 1997. The Impact of Information Technology in Sub-Saharan Africa with Particular Reference to Ethiopia. PhD. Thesis, University of Sheffield.

Apalayine, G.B., Ehikhamenor, F.A., 1996. The information needs and sources of primary health care workers in the Upper East Region of Ghana. J. Inf. Sci. 22 (5), 367–373.

Baldeh, Y.H.C., 1997. Information Support for District Health Care Planning and Decision Making in The Gambia: a Holistic Approach. PhD. Thesis. University of Central Lancashire.

Barton, T., Wamai, G., 1994. Equity and Vulnerability: A Situation Analysis of Women, Adolescents and Children in Uganda. Uganda National Council for Children, Kampala.

Belkin, N.J., 1978. Information concepts for information science. J. Doc. 34 (1), 55−85.

Braman, S., 1989. Defining information: an approach for policy makers. Telecomm. Policy 13 (3), 233−242.

Browne, M., 1997. The field of information policy: fundamental concepts. J. Inf. Sci. 23 (4), 261−275.

Chande-Mallya, R., 2014. Access and use of electronic health information resources among selected biomedical and health Universities in Tanzania. In: Proceedings of AHILA-14 Congress, Dar es Salaam.

Dervin, B., 1977. Useful theory for librarianship: communication, not information. Drexel Libr. Q. 13 (3), 16−32.

Dwivedi, Y.K., Wade, M.R., Schneberger, S.L., 2012. Information Systems Theory: Explaining and Predicting Our Digital Society. Springer, New York, NY.

Gadau, L.N., Lwoga, E.T., 2013. Information seeking behaviour of physicians in Tanzania. Inf. Dev. J. 29 (2), 172−182.

Gladwin, J., 1999. An Informational Approach to Health Management in Low Income Countries. Ph.D. Thesis, University of Sheffield.

Goel, S., et al., 2015. The health information seeking behaviour and needs of community health workers in Chandigarh in Northern India. Health Inf. Libr. J. 32 (2), 143−149.

Government of Uganda, 1997. Local Government Act. Ministry of Local Government, Kampala.

Justice, J., 1984. Can socio-cultural information improve health planning. Soc. Sci. Med. 19 (3), 193−198.

Kapiriri, L., Bondy, S.J., 2006. Health practitioners' and health planners' information needs and seeking behavior for decision making in Uganda. Int. J. Med. Inf. 75 (10−11), 714−721.

Kenney, G.I., 1995. The missing link−information. Inform. Technol. Dev. 6, 33−38.

Lemberger, P., Morel, M., 2012. Managing Complexity of Information Systems: The Value of Simplicity. ISTE and Wiley, London and Hoboken, NJ.

Lowan, B., Bukachi, F., Xavier, R., 1998. Health information in the developing world. Lancet 352 (Suppl.), 34−37, October.

Momodu, M., 2002. Information needs and information seeking behaviour of rural dwellers in Nigeria: a case study of Ekpoma in Esan West local Government area of Edo State, Nigeria. Libr. Rev. 51 (8), 406−410.

Musoke, M.G.N. (Ed.), 1997. The Uganda Health Information Digest 1 (1), ISSN: 1028−5105.

Musoke, M.G.N., 2001. Health information access and use in rural Uganda: an interaction-value model. PhD. Thesis, University of Sheffield.

Musoke, M.G.N., 2007a. Access to health information: the African users' perspective. In: Mlambo, A. (Ed.), African Scholarly Publishing Essays. African Books Collective Ltd, Oxford, ISBN:1-904855-83-0. pp. 120-128.

Musoke, M.G.N., 2007b. Information behaviour of primary health care providers in rural Uganda: an interaction-value model. J. Doc. 63 (3), 299−322.

Musoke, M.G.N., 2007c. Technology and repackaging enhance information delivery to remote Uganda. Paper presented at the FORO Conference, University of Arizona, February 2007.

Musoke, M.G.N., 2009. Document supply services enhance access to information resources in remote Uganda. Interlending Doc. Supply J. 37 (4), 171−176.

Musoke, M.G.N., Kinengyere, A.A., 2008. Changing strategies to enhance the usage of electronic resources among the academic community in Uganda with particular reference to Makerere University. In: Rosenberg, D. (Ed.), Evaluating Electronic Resource Programmes and Provision: Case Studies from Africa and Asia. INASP, Oxford, pp. 79−100, (Chapter 6).

Mutch, A., 1999. Information: a critical realist approach. In: Wilson, T.D., Allen, D.K. (Eds.), *Exploring the Contexts of Information Behaviour*. Proceedings of the Second International Conference on Research in Information Needs. Taylor Graham, London, Seeking and Use in Different Contexts, 13−15 August.

Namyalo, S., 2002. Challenges of Translating and Disseminating HIV/AIDS Messages in a Multilingual and Multicultural Nation: The Case of Uganda. Institute of Languages, Makerere University, Kampala.

Paek, H., et al., 2008. The contextual effects of gender norms, communication and social capital on family planning behaviors in Uganda: a multilevel approach. Health Educ. Behav. 35 (4), 461, http://heb.sagepub.com/cgi/content/abstract/35/4/461.

Patton, M.Q., 2002. Qualitative Evaluation and Research Methods. Sage Publications, London.

Rhine, L., Kanyengo, C., 2000. The development and use of the Guide to Medical Resources website at the University of Zambia medical library. Proceedings of the 8th International Congress on Medical Librarianship, July 2−5. Library Association, London, <http://www.icml.org/Tuesday/>.

Roberts, N., 1976. Social considerations towards a definition of information science. J. Doc. 32, 249−257.

Roberts, N., 1982. A search for Information man. Soc. Sci. Inf. Stud. 2, 93−104.

Ross, C.S., 1999. Finding without seeking: what readers say about the role of pleasure—reading as a source of information. In: Wilson, T.D., Allen, D.K. (Eds.), Exploring the Contexts of Information Behaviour. Proceedings of the Second International Conference on Research in Information Needs, Seeking and Use in Different Contexts, August 13−15. Taylor Graham, London.

Savolainen, R., 2015. Cognitive barriers to information seeking: a conceptual analysis. J. Inf. Sci. May. http://jis.sagepub.com/content/early/2015/05/27 (accessed 30.05.15).

Spink, A., 1999. Towards a theoretical framework for information retrieval in an information seeking context. In: Wilson, T.D., Allen, D.K. (Eds.), Exploring the Contexts of Information Behaviour, Proceedings of the 2nd International Conference on Research in Information Needs, Seeking and Use in Different Contexts, August 13−15. Taylor Graham, London.

Spink, A., Dee, C., 2007. Cognitive shifts related to interactive information retrieval. Online Inf. Rev. 31 (6), 845−860.

The Uganda Communications Commission. http://www.ucc.co.ug/data/qmenu/3/Facts-and-Figures.html (retrieved 23.04.15).

Uganda National Population and Housing Census Report, 2014. http://unstats.un.org/unsd/demographic/sources/census/2010_PHC/Uganda/UGA-2014-11.pdf.

Vakkari, P., 1999. Task complexity, problem structure and information actions: integrating studies on information seeking and retrieval. Inf. Process. Manage. 35, 819−837.

Vakkari, P., Järvelin, K., 2005. Explanation in information seeking and retrieval. In: Spink, A., Cole, C. (Eds.), New Directions in Cognitive Information Retrieval. Springer, Berlin, pp. 113−138.

Vakkari, P., Huuskonen, S., 2012. Search effort degrades search output but improves task outcome. J. Am. Soc. Inf. Sci. Technol. 63 (4), 657−670.

WHO, 1978. Primary Health Care. Report of the International Conference on Primary Health Care, September 6−12. Alma-Ata-USSR. The World Health Organisation, Geneva.

WHO, 2008. Primary Health Care: Now More Than Ever. The World Health Organisation (WHO), Geneva (The World Health Report 2008).

Wilson, T., 1981. On user studies and information needs. J. Doc. 37 (1), 3−15.

Wilson, T., 1997. Information behaviour: an interdisciplinary perspective. Inf. Process. Manage. 33 (4), 551−572.

Wilson, T., 1999. Exploring models of information behaviour: the uncertainty project. Inf. Process. Manage. 35, 839−849.

World Information Society, 2014. Measuring the Information Society Report.

ACCESS AND USE OF INFORMATION BY WOMEN AND HEALTH WORKERS

2

2.1 METHODOLOGICAL APPROACH

The main focus of the book is to gain insight regarding the perceptions and experiences of rural women and health workers' needs for health information, the actual and potential sources of information, the best and easiest ways of accessing information in rural areas, the various methods of information delivery to rural Uganda, the factors affecting access to health information and the use of health information by women and health workers.

Information access and use as well as the associated processes are, therefore, presented in a holistic way from the qualitative data. It was imperative to use a holistic inductive paradigm to be able to advance our understanding of the information access and use by women and health workers in rural Uganda. This is because a holistic approach studies a phenomenon in its entirety without reducing it to some variables. Data are collected on individual occurrences of the phenomenon, after which the analysis leads to patterns, themes and categories emerging out of the data inductively rather than deductively.

In addition, a holistic approach to information access was used to discover the interdependencies and relationships between information sources, needs for and use of information as a whole phenomenon. Information-related behavior was also considered from a holistic perspective and is not solely limited to active seeking behavior.

As already indicated, an inductive strategy was adopted instead of logical deductions from set hypotheses. This was because quantitative measurements could not lead to an understanding of how people perceived, understood and interpreted the information they accessed, given that a number of these important issues could not easily be quantified. Through close and direct interaction with the people in an open-minded inquiry and inductive analysis, this book is able to shed light on the phenomenon of information access and use in rural Uganda.

Within the interpretive research, the type of qualitative analysis that was considered to be most compatible with a holistic inductive approach was Grounded theory. Among the qualitative analytical approaches, Grounded theory was chosen because of its ability to generate theoretical models systematically through the constant comparative method whereby data, emerging concepts, categories and their properties are constantly compared. Although the study had a holistic inductive perspective with a Grounded theory approach, it differed from the Grounded theory as originally defined by Glaser and Strauss (1967) in that it did not adopt a theoretical sampling strategy. However, some aspects of theoretical sampling were used. For example, initially data were

Informed and Healthy. DOI: http://dx.doi.org/10.1016/B978-0-12-804290-8.00002-6

collected and analyzed, and the interview schedule was modified to include concepts that emerged from the analysis and were considered important to the phenomenon undergoing study. Such concepts included the church/faith-based organizations and seminars (as sources of information), unmet needs for information (these were from the pilot study), and coping, which were added to the interview schedule, thereby enriching the conceptual scope of the study.

The sample was determined by the Primary Health Care (PHC) setup as outlined in Chapter One and followed a purposeful sampling strategy as described by Patton (2002). The study focused on two categories of people from the lowest level of PHC service delivery, namely, the women and health workers. The total sample comprised eighty-two (48 women leaders and 34 health workers) from Bushenyi, Iganga, Lira and Masaka districts, which represent the traditional four regions of Uganda, namely, Western, Eastern, Northern and Central, respectively.

A study of health information access in rural Uganda had to take into account the PHC provisions and the Uganda Ministry of Health (MoH) National policy guidelines, which indicated that PHC was to be supported equally as secondary and tertiary care, and the community's input to health was valued equally with that of hospitals and health centers. The centerpiece of this was the creation of health subdistricts in which both the community-based health care and the work of health centers and hospitals were to be coordinated from a single center.

In Uganda, the district was regarded as the most appropriate level for implementing PHC because it provided a suitable organizational framework within which to introduce and implement changes and reforms in the health system. Furthermore, it was at the district level that national programs were adapted to local conditions. Districts were selected as study sites.

Furthermore, as indicated in Chapter One, the PHC within the Uganda national health policy recognizes the vital role women play in the health of their families and community. Hence, in an ordinary setting, it was first the women and then the health workers who provided health care. The study focused on the two groups of health providers: women and health workers in rural areas within a district. In addition, institutions that provided information to women and/or health workers were approached for complementary documentary evidence.

Information-rich cases among rural women in a subcounty were selected from grassroots women leaders, namely, the executive committee members of the local authority councils (23), women councils (10), women groups/clubs (9) and faith-based organizations (6), who can also be referred to as "information gatekeepers." The strategy used to select the purposeful sample was that of "critical case sampling," which permitted logical generalization and maximum application of information to other cases. Critical case sampling strategy involved selecting (or sometimes avoiding) cases that were, for some reason, particularly important in the community; for example, the sample included women leaders of different groups/organizations to tap the different aspects of community life.

Health workers were also purposefully selected for interviews and the different categories were included in the study, namely: medical doctors (8); clinical officers (9); registered nurses/midwives (5); enrolled nurses/midwives (7); and nursing assistants and Traditional Birth Attendants (TBAs) (5). For both women and health workers, sampling and data collection stopped when similar responses were received repeatedly and no new categories were forthcoming; hence, data were considered to be saturated.

An interview guide was used initially to pilot the research instruments in one district. However, given that different interviewers who speak the local language were to be involved, the interviewer flexibility in wording and sequencing questions highlighted the shortcomings of that method. An interview guide method proved unsuitable for use in the main study. Consequently, a semi-structured and open-ended design evolved as the best alternative in that situation for the following reasons:

i. interviewees answered the same questions, which enhanced the comparability of responses (and this facilitated cross-case data analysis);

ii. the data were complete for each interviewee on the topics addressed in the interview;

iii. the open nature of questions allowed further probing into the responses that greatly enriched the data collected and

iv. due to the language problem, the study inevitably used research assistants to interview the women in the local languages; hence, a semi-structured but standardized open interview schedule proved very useful in minimizing interviewer effects and biases.

The effect of the interviewer's gender on the interview situation was taken into account, particularly when interviewing women. The research assistants who collected data from women were, therefore, female social scientists who had been involved in various qualitative studies. Two male research assistants collected data, with the author, from TBAs. The professional experience of research assistants was a great asset to the data collection process.

The interview schedule was designed in English, and the health workers were interviewed in English, except the TBAs. Given that the women and the TBAs were not comfortable with the English language; therefore, the interview schedule had to be translated into the local languages and interviews were conducted in the vernacular. This was compounded by the fact that the word "information" does not exist in the vernacular, and Uganda has no national language, as already highlighted in Chapter One.

Face-to-face interviews were then conducted until the data were saturated (and no new categories were reported). Two sets of interview schedules were used to collect primary data: one for the women and the other for health workers. The health workers' interview schedule was slightly longer than that for the women because it included questions about the provision of health information to women and about health workers' professional contacts and sources. Both sets of the interview schedule focused on the two major issues the study set out to investigate, namely: accessibility and the use of information. These led to four subgroups. Under accessibility were the need for information, sources of information and factors affecting access; whereas under use of information, interviewees were asked about the use(s) for the information they accessed, because this had to be identified or experienced by the interviewees themselves. The questions were designed to study the interviewees' experiences, behaviors, knowledge and opinions.

Furthermore, the interview schedule included a question on critical incidents. Interviewees were asked to focus on a single recent critical health information incident and to describe it in detail from the time the need for information was felt to the time the need was satisfied or the incident ended. This question enriched data on information needs because it added a critical aspect. It also highlighted sources of information in critical situations, information acts, or how one went about getting the needed information, what that information did or did not to satisfy a need and how it did or did not solve a health problem. Furthermore, the question shed light on the consequences of lack of appropriate and/or timely information. The critical incident technique had been recommended by several qualitative research authors such as Patton (2002), Bryant and Gray (2006) and

Hughes-Hassell and Agosto (2007) for being robust, focused and methodologically sound. Odini (1995) pointed out that people tend to have less difficulty in recalling accurately a critical incident of a particular type and that the critical incident technique was useful in identifying some important factors influencing information seeking and communication behavior. Furthermore, it was noted that the qualitative research interview method emphasizes the importance of reflective listening to encourage the interviewee to reflect on his or her experiences. Urquhart et al. (2003) pointed out that the critical incident technique is based on that principle.

Finally, personal data, such as age and education (and working experience in the case of health workers), as well as information about the infrastructure in the area were also recorded during the interview to provide some background information and to put the interview setting in context.

While in the field, both real-time notes and audio-recording were undertaken. During the analysis, one of the records was used as a back-up of the other, except for the critical incidents when the recorded and transcribed notes were the main record.

The descriptive data were quantified using simple summations and frequencies, but the bulk of the analysis was interpreted using a Grounded theory to enable the researcher to discover concepts and relationships in the raw data (which eventually emerged into a theoretical model). This provided insight and understanding of health information access and use in Uganda.

Indeed, Grounded theory is a method of analyzing qualitative data to discover theory from that data. Strauss and Corbin (1990) elaborated it further by stating that the Grounded theory approach is a method that uses a systematic set of procedures to develop an inductively derived theory about a phenomenon. Data collection, analysis and eventual theory are in close relation to one another. Grounded theory data analysis was, therefore, compatible with the holistic inductive paradigm used in the study.

Given that a semi-structured interview method was used to collect data, so it was preferred to perform a cross-case analysis for each question in the interview schedule, which involved grouping together and comparing answers from different interviewees to common questions. The analyst then abstracted from the data and generated concepts and categories inductively. This was preceded by open and selective coding based on the original Glaser and Strauss (1967) approach. The analysis closely followed the stages highlighted by Turner (1981) and Ellis (1993), which were based on the constant comparative method of analysis that is central to the Grounded theory approach.

During open coding, there were constant comparisons, as well as sorting and grouping of concepts, categories and the incidents that resulted in these categories. The constant comparative method, according to Glaser and Strauss (1967), has four stages: comparing incidents applicable to each category; integrating categories and their properties; delimiting the theory and writing the theory. Although this method of generating theory is a continuously growing process, each stage after a time was transformed into the next, earlier stages remained in operation simultaneously throughout the analysis and each provided continuous development to its successive stage until the analysis ended. This meant that the core and main categories emerged after comparison with a number of categories, and their subcategories, properties and incidents became clearer.

Some interviewee responses are given in this book. The responses are based on verbatim records (translated from the vernacular to English in the case of women) and are shown in the shaded rectangular boxes. For confidentiality reasons, each quotation indicates the interviewee number and the district rather than the name.

The design of the study is summarized in Fig. 2.1.

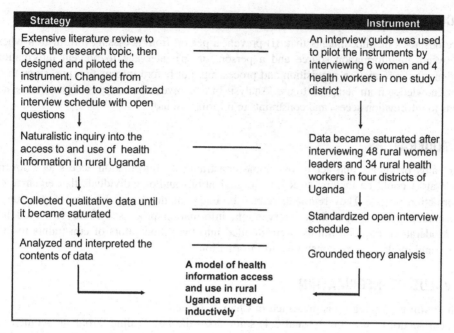

Strategy		Instrument
Extensive literature review to focus the research topic, then designed and piloted the instrument. Changed from interview guide to standardized interview schedule with open questions ↓	———	An interview guide was used to pilot the instruments by interviewing 6 women and 4 health workers in one study district ↓
Naturalistic inquiry into the access to and use of health information in rural Uganda ↓	———	Data became saturated after interviewing 48 rural women leaders and 34 rural health workers in four districts of Uganda ↓
Collected qualitative data until it became saturated ↓	———	Standardized open interview schedule ↓
Analyzed and interpreted the contents of data		Grounded theory analysis

A model of health information access and use in rural Uganda emerged inductively

FIGURE 2.1

Study design.

Musoke, M.G.N., 2001. Health Information Access and Use in Rural Uganda: An Interaction-Value Model. PhD. Thesis, University of Sheffield.

2.2 THE CATEGORIES

There were two "root" and three "emergent" categories that formed five preliminary categories: information sources; information needs; constraints; moderators and value of information. The last three categories emerged through Grounded theory analysis; hence, "emergent" as opposed to "root" categories, which the researcher started with. The root categories, namely information sources and needs, were identified from literature review, but their detailed findings were derived inductively from the data. The categories are outlined in this section.

2.2.1 INFORMATION SOURCES

This is where the information was obtained from (actual) or could be obtained from (potential). Information sources exist even when there is no apparent (active) need for information on the part of an individual.

2.2.2 INFORMATION NEEDS

The apparent need for information makes people go to the source to seek information actively, whereas for latent needs, people get information passively and then may realize that they needed it.

2.2.3 CONSTRAINTS

These are obstacles or impediments that: (i) prevent a person from accessing information when they occur between an information source and a person, or (ii) intervene to prevent information use. Intervention after information acquisition and processing, but before information use, stops the information or knowledge from being put to use. Analysis of data revealed two types of constraints, namely, constraints to information access and constraints to information use.

2.2.4 MODERATORS

These are aids or agents that act to overcome constraints to information access and information use. Moderators could be factors, structures, organizations and/or individuals that enhance or support information access. They regulate, reduce or intercept the constraints to information access and information use. Without their intervention, information processes could be halted by constraints. Moderators, like constraints, were divided into two: moderators of constraints to information access and moderators of constraints to information use.

2.2.5 VALUE OF INFORMATION

The fifth preliminary category is presented in Chapter Three.

The next sections (Sections 2.3 and 2.4) summarize the key findings from the women and the health workers and present them using the aforementioned categories.

2.3 INFORMATION ACTIVITIES BY WOMEN

The main analytical approach in the book followed an interpretative method, as already indicated, whereas the quantifiable data were supplementary and are presented first to set the scene in the different categories. Furthermore, authors such as Glaser (1978) and Strauss and Corbin (1998) emphasized that the discipline and rigor of qualitative analysis depend on presenting solid descriptive data, often called "thick description," in such a way that others reading the results can understand and draw their own interpretations. The rest of this chapter attempts to do that, starting with the information sources category.

2.3.1 INFORMATION SOURCES

This section includes actual and potential sources as well as information behavior.

Actual sources of information

Table 2.1 shows the actual sources of health information in order of importance. Women reported as many health information sources as they could easily remember. The number of times an information source was mentioned were added together to show its effectiveness; the first listed source was mentioned the highest number of times, and the reverse is true for the source listed last in Table 2.1.

Among the formal sources used to access health information, radio scored highest, followed by health workers, and books and pamphlets emerged last. As an information medium in rural Uganda, radio seemed to have evolved beyond its capacity to entertain; the relative affordability of radio sets and batteries allowed it to play a primary role in the provision of health information.

Table 2.1 Actual Sources of Information for Women

Actual Channels/Sources of Information

1. Radio	10. Other NGOs[d]
2. Health workers[a]	11. School children
3. Seminars[b]/workshops/outreach	12. Newspapers
4. Church/faith-based institutions	13. Television[e]
5. Friends/village-mates/relatives	14. Posters
6. Mobile phones	15. Traditional herbalists/TBAs
7. Drama/songs/poems	16. Films/video
8. Local councils[c] (LCs)	17. Books/pamphlets
9. Women's formal NGOs or informal clubs/groups	

[a]Health workers included those based in health units, but also health visitors, educators and inspectors from the District Medical Office, MoH and NGOs, but excluding traditional birth attendants (TBAs). Health workers also provided information on mobile phones.
[b]Seminars/workshops/outreach sessions were generally conducted by health workers; however, some, for example, those supported by the Elsevier Foundation, DFID, etc., were conducted by information professionals and health workers. A few political education seminars had a section on health.
[c]The local council/authority from the village (LC1) to the district (LC5) provided information even to the LC members themselves, usually with the guidance of health workers.
[d]Other NGOs excluding women's, for example, the Red Cross, The AIDS Support Organization (TASO) and other AIDS NGOs, Care International, Family Planning Association, UNICEF, Action Aid, Water and Sanitation and several community-based health NGOs or CBOs.
[e]TV: One of the public health messages on TV was about polio immunization, "Kick polio out of Uganda." Several interviewees quoted the TV as a source of information on polio immunization.

It was also noted that low education or lack of education tended to encourage an oral information culture among the women interviewed. This was compounded by the general lack of libraries or information centers in the rural areas studied. Hence, audio information scored highest, whereas books and pamphlets scored lowest as actual sources of information.

Another issue that Table 2.1 raises is the objectivity or overdependence on numbers in quantifiable data. If a sensitivity test is performed on the listed sources and, for example, radio or seminars were withdrawn as sources of information, then they would seem to have far greater effects on information access than if books and pamphlets were withdrawn. However, given that seminar facilitators, tutors and other information providers get most of their information from books, in reality, withdrawing printed sources from a Ugandan information scene would bring many information processes to a halt.

Sources that performed an information function, in addition to a primary noninformation function, were the local councils (LCs) and the faith-based organizations concerned primarily with administration and religious matters, respectively. They were both ranked high as sources or channels through which health information could be accessed. Other channels and sources of information involved people such as health workers, women in women's NGOs/groups/clubs, traditional herbalists, school children and friends/village-mates/relatives. Drama also fared quite well in the provision of health information to rural areas.

These findings generally agree with those reported by Muthoni and Miller (2010) in their study on breast cancer in Kenya. They pointed out that women would be reached best through community-based structures such as health facilities, women's groups, churches and family and friends. Furthermore, Momodu (2002) reported sources of information in a rural Nigerian setting, such as radio, television, extension workers, teachers, friends and relatives.

This literature shows that the sources of health information for women in rural Sub-Saharan African had generally remained the same in the past decade. However, although the 2014/15 follow-up study conducted in Uganda by the author identified similar sources of health information, there were more health programs on radio and TV presented by medical professionals and focusing on various health topics and more people had radio and TV sets than before.

Furthermore, according to the interviewees, one of the recent major changes on the information scene was the individual and/or family ownership and use of mobile phones. That channel made it easy to communicate because more people had mobile phones than a decade ago, when even some health professionals did not have phones. People are able to consult health professionals by phone, and they (as local leaders) also received health messages from various providers, including the MoH.

Potential sources of information

Other sources of health information were identified by the women as ones that they knew of or heard about, but they had not used or received information from them yet. Some pointed out that they had not fully exploited the sources that made them remain as potential or unused sources. They included:

  i. the Internet;
  ii. domestic NGOs with branches in different districts that run seminars/workshops that women had missed;
  iii. international NGOs operating in some rural areas but not others;
  iv. health-related Government departments and programs that did not reach some rural areas;
  v. formal sources such as libraries, which were reported to be located in towns and, therefore, generally inaccessible to rural people. Printed sources such as magazines, newspapers and pamphlets were also reported as inaccessible by some interviewees.

In summary, 58% (or 28) women identified these as potential sources. However, 42% of the interviewees reported that they had used some of these sources, although not fully. It was also noted that some of the sources that were reported as potential had been reported as actual sources by some women interviewees who had benefited from their information activities. The Internet remained a potential source of information to many people in Sub-Saharan Africa. The section on potential sources, therefore, shed some light on women's information needs and constraints affecting information sources.

Best or easiest ways to access health information

Women pointed out that what they perceived or experienced as the best ways/channels to deliver health information to rural areas may not be the easiest and vice versa. Radio was ranked as the best ways/channels, followed by health workers, and the easiest ways/channels were LC authority, mobile phones, faith-based functions/leaders, women groups and women leaders, drama/school children, social gatherings, respected elders and posters (in that order).

Based on their experiences, health workers were also asked to report on what they considered to be the best and easiest ways to deliver information to rural women. The majority (26 or 76%) of the health workers interviewed reported that health workers (either directly or indirectly through radio, mobile phones or other media) were the best source or channel because they would deliver professional and authentic information. Regular health education sessions, seminars, meetings and rural outreaches at village, parish or subcounty levels were considered the best way to deliver information, whereas some reported that house-to-house visits would be the best because they would ensure that all homes were reached, information was delivered, questions were answered and/or advice on various health aspects was given according to the situation in the homes. The human and material resources, however, would be the challenge. Consequently, the delivery of information by a health worker to a group of people in the villages was more practical and cost-effective.

Other analytical subdivisions

Further analytical insights of information sources revealed three subdivisions, namely, preference, information behavior (discussed in this section) and constraints concerning information sources (discussed in the Constraints category) (Section 2.3.3).

Preference

The choice of the information source was influenced by the user's personal attributes such as emotions, negative beliefs and attitudes, the type of illness, the type of information needed, and mainly by the characteristics of information source, which included accessibility and the user's perceptions about the credibility or reliability of information, as well as its reciprocity or interactivity. Lack of reciprocity or providing feedback that constrained information access is presented in Section 2.3.3 (Constraints).

Reliability of quality and accuracy, authenticity, trust or credibility of information affected the choice of information source. Wilson (1997: 561) observed that "if a seeker of information discovers that an information source is unreliable in the quality and accuracy of the information delivered, he or she is likely to regard the source as lacking in credibility." Women pointed out repeatedly that health workers had superior knowledge and therefore were a preferred source of credible information, although some lamented about their unavailability. It was also noted that women who could access the appropriate level of books and were able to read found books to be a reliable source of information.

To a lesser extent, women as leaders and information providers pointed out that the women they mobilize tend to think that some sources or providers had become monotonous. So, something new or a different provider was considered to be more credible, as the following comment shows:

> When we call meetings as women council, women ask us: is there anything new you are going to tell us in that meeting — and we try to provide something new but most of the time we have to follow up on our previous work ... So, we find that the turn up is poor. However, when we tell them that we have a visitor coming from outside, the turn up is very good
>
> **Woman 1, Iganga**

Women preferred sources that were easy to access. Although as already pointed out, credibility or reliability was considered important, as was accessibility, according to the majority of women interviewed. For example, friends and relatives were considered the least credible sources by some women who referred to the information they provided as "hearsay"; however, the inaccessibility of credible sources left some women with hardly any options but to rely on those who were easily accessible. Accessibility, as reported by the women, included physical location of the information source and, in the case of human sources, a combination of physical location and willingness to provide the needed information. For example, LC officials and TBAs/old women were reported to be more accessible than health workers. Similarly, friends, neighbors and drug shops were reportedly more accessible than health workers/health units.

> Usually when I need advice on health issues, the first person I consult is an old woman or a TBA because they are available, easily accessible, and they willingly provide the information although

> sometimes, they also advise me to consult health workers whom one has to look for at the sub county health centre, which is costly... I only consult them on major issues which other people can't handle... otherwise, for obvious illnesses, I go to a nearby drug shop and buy medicine
>
> **Woman 1, Lira**

Although Tipping and Segall (1995) reported that the reason the literature they reviewed suggested preferring over-the-counter drugs or treatments (OCTs) was because people were avoiding health unit queues, these findings highlight that accessibility of OCTs is another issue. To many rural people, health units were far away, which made them prefer going to a nearby drug shop to get advice and/or medicine, particularly if it was not the type of consultation one could make by phone.

Besides being physically accessible, the church or places of worship in general were reported to be socially acceptable, and women had no problems with their spouses when they attended religious functions. This made women identify the church as an easy and accessible source or channel of information. Generally, there were places of worship at every parish, and yet health units were at the subcounty level.

To many interviewees, radio was more accessible than health workers or health units. Radio also catered to other access factors, such as illiteracy and heavy workload. Such problems tended to dictate the choice of information source. Hence, radio, which provided information to people while doing other tasks and chores at home, was the best choice for many as indicated in the following comment:

> Radio reaches all of us and smaller radio sets are affordable... I like the radio because I can listen while doing my other work unlike TV which has to be watched... those who can't read or who don't access printed sources do benefit from the radio
>
> **Woman 4, Masaka**

Reciprocity that facilitates information exchange, getting feedback, and/or interaction between the source and the recipient was another important characteristic of an information source. Women preferred health workers/units because they would ask questions and receive answers, unlike radio that some criticized for not being able to get feedback. The term "reciprocity" is loosely used for the type of information exchange referred to in the following comment:

> The advantages and disadvantages of different family planning methods are not included in the messages I have heard on radio. Such information is provided in Family Planning clinics, which are a very important source of information because one gets an opportunity to ask health workers questions...and get a response...
>
> **Woman 2, Bushenyi**

These findings are well supported in the information science literature. Johnson and Meischke (1991), for example, found that interpersonal sources are better suited to handle special individual needs and questions due to the immediate feedback available from the source. They further pointed out that interpersonal sources, such as consultations with a health worker, can be more effective in reducing uncertainty for cancer patients, for example, because they provide immediate feedback and social support. Both of these factors give the patient confidence in the advice received.

The type of illness sometimes dictates which sources to consult and which ones to avoid. Some diseases or illnesses were classified by the women as sensitive, usual or unusual, whereas others were of a general nature. Some women pointed out that unless it was what they referred to as an ordinary disease, one could not just look for information from anywhere or discuss the problem with anybody. Some diseases reportedly required privacy, and professional health workers were preferred to friends or neighbors. However, in some cases, the sex and age of health workers were also issues; hence, in the absence of female health workers, women preferred discussing what they perceived as sensitive or private health issues with their fellow women, as illustrated in the following comment:

> It is difficult to ask neighbours or friends about some sensitive issues such as AIDS, one has to confide in health workers...otherwise, neighbours can gossip about one's problems (woman4-Bushenyi); We have a problem here because all the nearby health units are run by youngish male health workers only... many of us find it difficult to discuss some personal diseases such as venereal wounds with young males... So, we seek advice from fellow women or go to the district hospital
>
> **Woman 3, Iganga**

In some situations, health workers were also the preferred choice when the illness was of the usual type, such as malaria, in which case one would seek treatment from a health worker rather than information. In other situations when the illness was perceived as unusual, some women preferred consulting health workers, whereas others preferred consulting their social networks first. Even for general diseases or health problems, which were neither sensitive nor unusual, the source of information depended on the illness, for example:

> It depends, if the problem is related to child birth, I consult an elderly woman or TBA; in other cases, I go to the dispensary and consult a health worker (woman2-Bushenyi); issues concerning 'false teeth' and traditional practices during pregnancy, I consult TBAs... Some health problems which require counselling or spiritual help, I consult a priest or other religious people
>
> **Woman 1, Lira**

Women's perceptions of health issues or sickness play a role in their choice of information source, and the source tends to vary according to women's beliefs, emotions and/or attitude as illustrated:

> To find out more about the effects of taking contraceptive pills for a long time, I will consult a doctor not these people who distribute pills... I think they wouldn't tell me the truth because their job is to give pills... if they tell me the negative effects and I stop taking pills, they may have no job
>
> **Woman 4, Masaka**

The type of information needed determined the choice of information source. When women needed factual or recorded information, they had to go to the source of the record as indicated here:

> If the information I need is, for example, the date or venue of immunisation, I go and check at the poster... but when I misplace the labels which show the doses of the medicine I am given, I have to go back to the health unit where I got it from
>
> **Woman 6, Lira**

Information behavior

Another issue that emerged while analyzing the data about information sources was information behavior. It also is discussed in other sections of the book. What is discussed here is focused on sources.

Passive versus active access to health information

As indicated in Table 2.1, apart from attending seminars/workshops and occasional reading of printed sources such as seminar/workshop handouts, books/pamphlets and newspapers, the majority of the information was accessed passively, for example, by listening to radio, preachers in church or mosque, health workers and LC officials, receiving a message on a mobile phone, watching TV, and through conversations with friends, relatives and village-mates. Listening and watching are natural senses, whereas reading (of printed sources or text message on phone) has to be acquired. Although there was generally limited availability of simplified printed health information in the rural areas studied, there was also a problem of poor reading culture, even among the literates. Passive access to oral or audio information was, therefore, predominant.

Hence, information acquisition can be categorized using sensory attributes such as listening (to the radio, health workers, LCs, preachers, songs, school children, etc.), watching (drama, television, films/video, etc.) and talking (with friends, relatives, etc.), thus further showing the predominance of passive access over active seeking of health information by the women interviewed.

In critical situations, however, women actively sought information mainly from relatives[1]/friends/neighbors and health workers. This slightly differed from the work of Phillips and Zorn (1994), who found that consumers "overwhelmingly indicated their personal physician or other health professional as the first point of call." The difference may be attributed to the situation in rural Uganda, where such a thing as "personal physician" rarely applies. In the findings, many interviewees (95%) indicated that they started with relatives, friends and/or neighbors, and then moved on to health workers.

Use of radio, television and newspapers in active information seeking

Although Wilson (1997: 562) refers to listening to the radio as "passive attention," and indeed many (75%) women in the study accessed information passively from radio, 25% pointed out that they deliberately switched on the radio to listen to specific health programs. Some read newspapers to check relevant TV and radio programs, whereas others used newspapers to get current information about health from the "health" page. Such selectivity in media use is, therefore, based on interest, which agrees with the work of Severin and Tankard (1988), who reported that "Many viewers 'actively' choose between competing newscasters, arrange their schedules to be near a television set at news time, and pay close, albeit selective, attention to the program" (p. 304).

Hence, the findings show that women actively chose between competing mass media programs, as well as competing demands on their time, and decided to listen to, watch, or read about health programs that they expect to address their information needs.

2.3.2 INFORMATION NEEDS

This section includes what the interviewees perceived and reported as their needs for health information, that is, "what they needed to know" and, to some extent, "what the information was needed for." Such perceptions confirmed the subjective[2] nature of information needs as well as the voices of the information user in this book. The notion of information needs was reported by some authors (Wilson, 1981) to be an abstraction derived from other human problems or needs. In the case of this book, it refers to health needs; therefore, information was needed to satisfy some human needs regarding health.

---

[1]Relatives mentioned were husband, sister and mother-in-law, in that order.

[2]"Need" is a subjective experience that occurs only in the mind of the person in need and, consequently, is not directly accessible to an observer. The experience of need can only be discovered by deduction from behavior or through the reports of the person in need (Wilson, 1997: 552); however, a physiological need such as hunger can be observed.

All the women interviewed reported that they had needs for health information. Some of their needs were met, whereas others remained unmet. Given their different roles, women needed information as PHC providers in the family, as leaders to cater to community health needs, and for personal reasons.

Information that was not accessed

Information that had not been accessed by the rural women interviewed, what made women need information frequently, and the health topics or diseases that women needed information about are summarized in Table 2.2 in order of importance, with the first having been mentioned the highest number of times.

The need for information about malaria was supported by the fact that malaria was the leading cause of morbidity and mortality in Uganda, a country bearing the third-largest burden of disease in Africa in 2014, where malaria affected mainly children younger than five years, as reported by Ndira et al. (2014). It was further noted that in the Bushenyi district, malaria in children was the major need for information and no interviewee mentioned diarrhea; however, in Lira district, 50% of the women highlighted information on diarrhea as a need, in addition to malaria. In the Iganga district, the major childhood health problems were measles and malaria, and a few mentioned diarrhea; however, women in Masaka highlighted child health in general and some singled out measles, but only one mentioned diarrhea. Furthermore, whereas interviewees in Bushenyi, Masaka and Iganga districts reported that information about cholera, hygiene and sanitation was satisfactory, in Lira some women pointed out that the information they had received was incomplete and they still needed more.

Further analysis of data indicated that the information that women needed but had not been able to access was mainly in regard to the causes of diseases/health problems and how they could be avoided; hence, the information needs were more preventative than curative. However, with regards to the health issues for which some information had been accessed but more was needed, women had various information needs ranging from causes of diseases/illnesses to making choices about health, which were both curative and preventative needs. The various needs for health information and the number of women who identified them have been summarized in Table 2.3 using the concepts that emerged from the data and arranged in order of importance, where number one scored highest.

Table 2.2 Health Topics/Diseases for Which Women Needed Information

Topics That Women Needed More Information About

1. Childhood diseases, including malaria	11. Asthma
2. HIV/AIDS	12. Skin diseases and leprosy
3. Reproductive health, including family planning (FP), infertility, pregnancy complications	13. Backache and other joint pains
	14. Headaches, epilepsy, meningitis and mental illness
4. Sexually transmitted diseases/sexually transmitted infections	15. Cancer
	16. Water-borne diseases
5. Malaria, general	17. Diabetes
6. General health and nutrition	18. Eye diseases
7. Heart diseases	19. Tetanus
8. Ulcers	20. Stomachaches, hernia and related illnesses
9. Medication/medicines	21. Ear, nose and throat diseases
10. Sickle cell disease	22. Adolescent health
	23. Elephantiasis

Table 2.3 Women's Needs for Health Information

What Women Needed to Know	
1. Causes of disease(s)	8. Relationship between diseases
2. Treatment/resistant diseases	9. Updating/health knowledge
3. Prevention of diseases/health problems	10. Coping with illness/health problem
4. Detection of diseases/illness	11. Overcome misconceptions
5. Community support	12. Overcome constraints
6. Home care/safety of medicines	13. Make choices/health decisions
7. Effectiveness of health program(s)	

Table 2.4 The Most Difficult Type of Information to Access

Most Difficult Type of Information to Access	Number/%
Printed information	34/71%
Professional health information/advice	9/19%
Health information in the vernacular	5/10%
Total number of interviewees/percent	48/100%

All women interviewees reported that access to health information in rural areas was not easy, and their comments show a relationship between information needs and constraints to information access, for example:

> In remote areas where health units are far and the LCs are not vigilant, it is very difficult to access information
>
> **Woman 1, Bushenyi.**

From the women's experiences, the most difficult type of health information to access was printed information. Women reported that they needed simplified printed information on health but that they had not been able to access it. This shows a linkage between needs, sources and constraints. Furthermore, it was noted that although health workers were second in the overall totals of actual sources of information reported by the women (Table 2.1), some women also pointed out that professional health information or advice was the most difficult to get. The quantifiable data summarized in Tables 2.2–2.4 have shown that, generally, less information trickled down to rural areas than was needed and women reported unanswered questions or unmet information needs.

Information that was accessed

Although there were still unmet needs for health information, women pointed out that they had accessed some information quite easily, as summarized in Table 2.5. It was noted, however, that some women had reported similar topics in Table 2.2 and that they needed more information about them. That points to the differences in the ability of the interviewees to access information.

Table 2.5 shows information about various health topics or diseases that had been accessed by women in the past 3 years. Women reported as many health topics/diseases as they could easily remember. The topic that scored highest is listed first, and the one with the least score is the last in the table. The scoring reflects some measure of awareness about the disease or health topic: the greater the total, the more awareness there had been about a topic or disease. It should be noted, however, that some women had reported similar topics in Table 2.2 and that they needed more

Table 2.5 Information Accessed

Health Topics/Diseases

1. HIV/AIDS and STIs (including condom use)	12. Influenza/coughs and tuberculosis
2. Hygiene and sanitation	13. Alcoholism
3. Immunization	14. First aid/snake bite, burns, wounds
4. Cholera	15. Disability
5. Some childhood illnesses	16. Asthma
6. Malaria and antimalarial drugs	17. Ulcers
7. Family planning (FP)	18. Sickle cell disease
8. Nutrition/balanced diet	19. Dental hygiene
9. Reproductive health (other than FP)	20. Diabetes
10. Adolescent health	21. Hypertension
11. Medicines/drugs	

Table 2.6 Overall Rating of Comments About the Information Accessed

Women's Perceptions of the Information Accessed	%
Satisfactory/good/complete/informative/educative	59.6
Fairly good, but some information was missing/lacked details	25.9
Incomplete/not satisfactory/left questions unanswered	14.5

information about them. It was not possible to have all the interviewees in the different parts of Uganda accessing health information at the same rate, and that is why there is a discrepancy.

Within the districts, Masaka and Lira had greater awareness of different health problems than Bushenyi and Iganga. This confirms that rural Uganda is not homogeneous; even within the same subcounty, some women leaders accessed information more easily than others. However, women's perceptions of the information so far accessed showed that some of their needs had not been fully satisfied due to a number of reasons, for example, some of the information they had received lacked detail, was incomplete or left some questions unanswered. An example of comments about HIV/AIDS, which had the highest awareness (see Table 2.5), indicated that, generally, interviewees were satisfied with the preventive and control messages, but information about several other aspects such as mother-to-child transmission, incubation period and relationship with other diseases was reported to have been unsatisfactory or not accessed at all. This may explain why HIV/AIDS is second in Table 2.2, which shows the topics for which more information was needed.

Table 2.6, therefore, shows that approximately 60% of the information accessed by women in the past 3 years was perceived to have been satisfactory and valuable; 26% was fairly good but some needs remained unmet; and approximately 14% left women's needs totally unmet. The reasons for lack of satisfaction point directly to constraints, thus further highlighting the relationship between unmet needs and constraints.

The analytical insights and interpretation of the data on information needs

Further analysis and interpretation of data on information needs highlighted three subdivisions: critical, active (but not critical) and latent. The corresponding information behavior in each of the three subdivisions was the other outcome of the analysis and interpretation of data, and it is presented as well.

The information needs can also be classified as met or unmet needs. Critical information needs led to active information seeking, so the majority were reported to have been met. However, some critical information needs were unmet, for example, when people failed to get appropriate or timely information that made the bad situation worse, sometimes ending in loss of lives. For active information needs, some were met but some were unmet, such as the ongoing need to cope with a disease or a health problem. However, latent needs were reported to be unmet. In summary, the subdivisions are as follows:

- Critical: majority met, some unmet
- Active: some met, some unmet
- Latent: generally unmet.

Critical information needs

In critical health situations, information was needed for interpretation of illness, treatment, decision making, prevention and community support. However, during the analysis of some of the interviews, there seemed to be a gray area between health information seeking behavior and treatment seeking behavior. Furthermore, due to the broad nature of the term "information need," it was noted that when interviewees narrated their critical health incidents, which made them seek information, they also included what purpose the information served and to what use it was actually put when received.

As far as the interpretation of illness was concerned, information was needed to know the type of illness, the disease, the cause of illness, and whether it is perceived as a scientific or traditional disease. It also includes, at a slightly lower level, confirming one's observation, opinion or belief. Many incidents were narrated about children's illnesses, for example:

> My one year old son has been getting frequent fevers which worried me… these days we fear frequent illness, you never know it could be AIDS. So, when he got the recent attack with convulsions, I asked some friends nearby what could be causing the problem. They thought it was malaria since many children had suffered from it; they advised me to take the child to a health unit as we were advised on radio '…in case symptoms persist, seek medical advice'. When I reached the health centre, I was told that the child had malaria; he was treated and got better, but since yesterday, he has not been fine… so, I don't know whether it is malaria again or something else
>
> **Woman 3, Bushenyi**

Furthermore, some of the incidents narrated revealed the consequences of delays. For example, delayed interpretation, on the part of women or health workers, coupled with lack of appropriate advice or information led to death in the critical situations reported by five women. In some cases there was conflicting information or advice and it was difficult to judge which one to use. This stresses the importance of timely and appropriate information. Provision "of the right information at the right time by the year 2000" was one of WHO's declarations, which seems to have remained on paper. In rural Uganda, it was not easy to access the right information or information source as indicated in the following example:

> I woke up one morning and realised that my baby was very weak and had high temperature; I thought it was 'omushuija'[3] (malaria). From about 8 a.m., I started asking village-mates for advice, what disease could it be, what do I do? One person kept on referring me to the next;

---

[3]In the vernacular (Runyankole), the word "omushuija" is used to refer to both fever and malaria.

> some said it was malaria, others that it was something else! By the afternoon, I decided to take the baby to a health centre but it was too late... the baby passed away!
>
> **Woman 4, Bushenyi**

It was noted that women's perceptions of illness or health problem and their preliminary interpretation of the illness play a role in their choice of source of information, and the source tends to vary according to women's own beliefs and knowledge about illness and the availability of relatives, friends or neighbors who become a primary source of information when they confirm the women's interpretation or assist in identifying and labeling the symptoms. This finding agrees with Bantebya-Kyomuhendo (1997), who pointed out that women's perceptions and interpretation of illness symptoms are entangled in the social support networks that seem to play a contributory role in the choice of treatment options. She further noted that there were often uncertainties in the women's assessment of illness that led to, among other things, the concurrent use of traditional and biomedical treatment.

An example of such incidents is highlighted here; it also takes us to the next critical need—management of patients:

> My first two children got measles — when I was expecting my third born — I consulted my neighbour who advised me to give them herbs but to stop giving them milk and meat... Two days later, the kids were not improving... My husband decided to take them to a clinical officer... who told him to stop giving them herbs and to give them a lot of protein — milk, meat, etc. When he told me, I sought advice from my neighbour again who insisted... I was really torn apart but my husband over- powered me and we followed the medical advice... the children got better
>
> **Woman 5, Masaka**

The use of both medical and traditional systems in a single illness episode has also been reported by medical anthropologists in modernization theories. They point out that the therapeutic choice remains pluralistic "irrespective of beliefs about causation, people tend to utilise 'modern' health services but would also consult traditional healers when the former fail to bring them relief... But it is clear from a range of evidence that indigenous African therapeutic systems continue to co-exist with Western bio medical practice throughout the continent, despite the fact that Western medicine enjoys considerably more legitimacy, public support, funding and status" (Wallman, 1996: 112, 3).

Given this, health workers recommended more health education of women to provide them with information that would facilitate early detection of illnesses and to emphasize the need to seek medical rather than traditional treatment as the following comment shows:

> Since it is usually women who first notice the health problem in a family, it is important to provide them with information to enable them to detect the diseases early; for example, problems such as fast breathing may be related to pneumonia; so, it is important to rush the patient to a health unit rather than wasting time with traditional herbs
>
> **Health worker 4, Bushenyi**

Information was also needed for treatment or management of children's, self and family members' illnesses or health problems. At a slightly lower level, there were information needs for "appropriate prescription/medicine," and there were needs for information concerning first aid, treatment or management of illness generally. It was noted that although people might have the means to buy medicine from nearby drug shops/pharmacies and treat themselves at home, they did not necessarily have the knowledge to administer the medicines. This was compounded by the limited knowledge of some medicines by drug shop/pharmacy attendants, who were sometimes reported not to provide appropriate information, thus losing trust or credibility from their clients who had preferred drug shops/pharmacies for being more accessible than health units. For example:

> I recently got medicine for my uterine pains from a drug shop but I did not trust the drug shop attendant who prescribed the medicine because I asked her some questions which she couldn't answer; so, I was not sure about the medicine. I then went to my neighbour and asked her whether she knew anything about that medicine but she didn't. She advised me to consult an elderly woman whom we always ask. When I reached there, she too did not know about that particular medicine. I got stuck... until I decided to go to the sub county health centre for advice. The Medical officer I found there decided to give me another type of medicine altogether, and I threw away the other one...time and money wasted!
>
> **Woman 2, Bushenyi**

However, the Uganda MoH is aware of such problems and has been vigilant in monitoring the operations of drug shops/pharmacies, and those that do not adhere to the guidelines get closed and licenses are withdrawn.

As expected, many critical incidents involved children. Several other incidents involved family members (husband, brother, mother, etc.). Women consulted neighbors, friends, relatives, health workers and/ or seminar notes to help them make decisions about treatment and other critical situations. For example:

> When my 12 year old daughter got bitten by a snake, I was not sure what to do! I called a neighbour for advice, who reminded me about a workshop[4] we had attended on first aid management of accidents including snake bites. I got out my seminar notes, followed the guidelines on first aid, together with my neighbour, after which we were advised to rush the affected person to a health unit... I then took my daughter to a dispensary... She recovered
>
> **Woman 7, Masaka**

Prevention and control of epidemics were other critical information needs that emerged during the analysis of data from Lira district in the main study. Unlike women in Bushenyi who reported that they had received satisfactory information that enabled them to prevent cholera in their area, some interviewees from Lira reported that they still needed information about cholera. Furthermore, in the follow-up study conducted in 2014/15, women reported that they needed regular updates of information to prevent epidemics such as Ebola from returning to Uganda based on the 2012 experience.

Community support was a critical information need initially reported in the Iganga district, and it demonstrates women leaders' roles in PHC, as narrated:

> Relatives of an AIDS patient at this village approached me recently for help when they realised that the patient's condition was deteriorating fast yet he had refused to be taken for treatment. They requested me to talk to the patient... I got my seminar note book and read about counselling of an AIDS patient; I followed the tips and guidelines noting each one of them carefully — I had forgotten much of it — then I went and approached the patient, counselled him and without much trouble, he accepted to go for treatment
>
> **Woman 3, Iganga**

It was observed that in critical situations, women's previous experiences and beliefs played a big role in their information behavior. Second, there were situations when women reported having acquired information previously without realizing that it was relevant to them personally, but they knew that it was relevant to their communities and therefore collected it. After some time, however,

---

[4]That was one of the Elsevier Foundation—supported outreach workshops in Masaka district.

the information turned out to be relevant and served them in critical situations, for example, those who referred to seminar notes to solve their personal, family or community information needs. This further confirms that women's information needs are dynamic as the following comment shows:

> My children had never suffered from serious diarrhoea; so, when I attended a seminar where we were taught how to mix ORS, I didn't take it very seriously. I took down notes just in case women ask me for advice as they always do. In fact, after the seminar, several women asked me and I gave them my seminar notes to read or I read out to those who can't read. However, a few months ago, one of my children got diarrhoea but I had forgotten the recommended quantities of mixing ORS. I got my seminar notes, read and mixed the ORS as recommended and gave to the child
>
> **Woman 1, Iganga**

Active information needs

These were mainly affective and cognitive needs, which led to some active information seeking from various sources. They were generally active but not critical needs. They included three major types of information needs: updating/health knowledge, community support and coping with illness or health problems. The first two were related to people's activities or responsibilities, interests and habits. Women's habits and interests, for example, reading newspapers, listening to the radio and watching the television were contributory factors to active seeking. Women leaders live and work in villages that are social settings, so the villages and the women they lead create their own demands that motivate women leaders to seek for health information from as many sources as they can access.

Health workers confirmed that there was demand for health education from the women:

> When we go for rural outreach programmes, women usually tell us that we should not start by giving vaccinations, that we should first health educate them and allow them to ask questions...'we want to learn' ('twagala misomo' in Luganda — Masaka; 'twenda kusoma' in Lusonga — Iganga, etc)
>
> **Health worker 4, Iganga; 3, Masaka**

Coping with illness or health problems was an active information need that became critical in some situations. Coping refers to active efforts to master, reduce or tolerate the demands created by stress. The need to tailor information to people has been expressed in the psychology literature. This is because information has been found to increase as well as alleviate stress. One of the first women interviewed in the first research district (Bushenyi) pointed out that she needed information *"to learn how to deal with the stigma associated with the AIDS problem... and the worries I have since my husband died"* (Woman 4, Bushenyi). The concept of "coping with stress," which fit that need, was particularly appealing to the researcher. This made her sensitive to the potential significance of this concept in subsequent interviews.

This book has shown that most interviewees needed information for problem-focused coping, whereas others, mainly persons living with AIDS or other life-threatening diseases such as sickle cell anemia, needed information for both emotion-focused and problem-focused coping,[5] as the following example shows:

> As a sickler, I need information to be able to understand a number of issues about the disease; for example, is it true that sicklers should not conceive because they might die during child birth, or produce sickly children... how come that some sicklers produce healthy children? I would wish

---

[5]The two major functions of coping are getting one's emotions (emotion-focused coping) under control and managing the problem (problem-focused coping) that has caused the distress (Baker, 1995: 70).

> to get a baby but I am scared! I also need information to be able to take better care of myself... what I should or should not do as this disease can take my life anytime; so, I always look for information about the disease, but I still have many unanswered questions... When I don't get the information I need, I get scared of death... Sometimes I get a lot of pain even when I am on treatment and I feel as though I am going to pass away, so I call a priest to say a prayer for me
>
> **Woman 5, Lira**

Most other information needs were for problem-focused coping, for example:

> My daughter gets frequent asthmatic attacks especially at night which makes me feel so desperate for help. I need to know how to prolong the girl's life and also to give her relief when she gets an attack. I have talked to many people without getting much help. Recently, I talked to one person with asthma, who advised me that when the girl gets an attack, I move her out in fresh air. When I do this, the girl regains herself until morning when we go to the hospital. But I also need to know how to control or reduce the attacks. Otherwise, I spend many sleepless nights
>
> **Woman 11, Iganga**

Women also needed information to update their health knowledge and to enable them to sensitize, inform or advise the community.

Latent information needs

As already indicated, many health information needs did not automatically lead to active information seeking. Women had needs that remained latent until exposed to information. However, some information needs were reported to have been active and to have led to active information seeking initially, but after failing to get what one needed, seeking did not continue (for various reasons, some of which are discussed under constraints to information access). Such needs, therefore, changed from being active to latent, which again shows the dynamic nature of information needs in the book.

Latent information needs included prevention, detection and causes of diseases, treatment/management of patients, the need to overcome sociocultural constraints, making choices, making health decisions and updating health knowledge.

With regard to prevention of diseases/health problems, the main concerns were immunization and, to a lesser extent, contraceptives, prevention of malaria and STIs. Many women pointed out that they wanted to get information about the effectiveness and safety of immunization. Some sought information actively, but others did not, and they simply decided not to take their children for immunization, thereby leaving this as a latent need. This issue, therefore, appears as both passive and active seeking, and it is also one of the issues this book identified as changing women's behavior from passive to active.

There were needs for information about the effectiveness of medicine/health programs and preventive measures in general, as indicated in the following comments:

> I need to know why children are immunised many times against polio; is it because the vaccine is weak... I heard it is called a booster dose, is it because they want to boost the weak doses? Secondly, how effective is BCG; why do some children get TB after being vaccinated? I also need to know why there is no vaccine for malaria, and why do some people get it repeatedly even after taking preventive measures?
>
> **Woman 10, Masaka**

Furthermore, women needed information about the safety of medicines. Although this is more on the preventive side, it has some aspects of treatment. The main concerns were also about immunization, medicines and contraceptives, for example:

> drugs like viagra have been advertised so much and people have used it but we have heard of fatal cases... However, there is no explanation from the MoH or other authorities whether people should stop or continue using viagra, and whether the drug can result into death!
>
> **Woman 7, Masaka**

> I am not sure whether FP pills don't bring abnormalities in children born after using them. My sister in law first used a coil (copper T) method, then she bled very much, and changed to pills. Last week she delivered a baby with a big swelling on the back which is scaring; In my case, when I decided to stop child birth, I went for tubal ligation, and I am fine. But my concern is for young women who need child spacing... what advice can we give them after these incidents? So, we need information about the safety of pills and other methods
>
> **Woman 10, Iganga**

There were many other general preventive information needs concerning childhood illnesses (eg, diarrhea, colds, tonsillitis, malaria, false teeth and mumps) and the prevention of various other diseases such as hypertension, uterine and breast cancer, meningitis, epilepsy, food poisoning and skin diseases. Most of the general preventive information needs had concerns about causes, but "cause" was a means to the end, which was prevention, for example: "...*what causes mumps so that I can ... protect my children...*" (Woman 12, Lira).

With regard to the detection of diseases/health problems, women needed to know the symptoms of infertility, cancer, yaws, elephantiasis, diabetes, TB, STIs and others, and how to detect the various diseases/health problems in early stages, as illustrated in the following comment:

> is it possible for one to know in advance that she is infertile so that she does not get married to be harassed for failing to produce children... I need the information to answer the women who consult me
>
> **Woman 9, Lira**

There were also information needs about causes of diseases and the relationship between them, for example:

> I need to know what causes hypertension... why do hypertensive patients also suffer from diabetes at the same time? (woman5-Lira); I would like to know the difference between cholera and similar diseases such as diarrhoea and dysentery (woman4-Lira); my father has had hernia problems repeatedly and so far he has been operated 3 times ... we don't know why ... I would also like to know whether hernia is hereditary — am I likely to get it too?
>
> **Woman 3, Masaka**

Treatment information needs included management of patients in general, as well as effectiveness of treatment and drug-resistant or persistent diseases. The most common concerns were regarding malaria and STIs; there were also concerns about the treatment of cholera, ulcers, epilepsy, goiter, impotence and elephantiasis. Some women needed information about the medical management of diseases such as measles because they knew only the traditional methods. Others needed information about specialized treatment and information sources because the nearby health units had not solved their problems. Other needs related to treatment were about self-medication, home care/home nursing skills and first aid. For example:

> I would like to get information concerning care for an epileptic patient, for example, how does one know that s/he is about to get an attack so that we can prevent her/him from falling and getting hurt?
>
> **Woman 3, Lira**

Health knowledge and capacity building information needs were mainly cognitive. Furthermore, information was needed to enable women to, for example, overcome constraints or misconceptions to be able to convince their spouses to use contraceptives. It was reported under prevention that some women needed information about the safety and/or the effectiveness of contraceptives. The needs were different here because these particular women had decided to use contraceptives, but their spouses did not support the idea. Consequently, this led to the need to overcome gender constraints as illustrated here:

> I need to know how to convince my husband that it is safe to use a condom... (woman3-Bushenyi); I wish I could know what to do... I think I need to learn more about the advantages of FP so that I can share with my husband and convince him to let me use contraceptives
>
> **Woman 8, Lira**

Other women needed information to enable them to overcome misconceptions:

> I would like to get information to confirm that a condom is comfortable, can't burst or get torn during the act
>
> **Woman 7, Iganga**

Furthermore, information was needed to enable women to make choices, for example, between different family planning (FP) methods or make decisions to start FP. In most cases, this was related to overcoming misconceptions and constraints. Such information needs were confirmed and justified by health workers who lamented the limited health facilities, especially FP services, at the grassroots, for example:

> The Village Health Team[6] members who also act as FP community based distribution agents at grassroots level are few and get overstretched... The fact that FP services are minimal at the grassroots and the available information is incomplete have led to rumours and misconceptions about FP; e.g. 'pills cause cancer, injections cause blood pressure, condoms can get stuck into the woman and kill her, vasectomy causes impotence.' Although FP programmes have tried to sensitise lower level health workers and communities, a lot still needs to be done. Government has put comparatively less amount of money in reproductive health than programmes such as cholera and malaria control, which is understandable
>
> **Health worker 6, Bushenyi**

Information behavior

Two broad areas emerged from the data: behavior related to coping and general information behavior that was subdivided into active and passive.

---

[6]The Village Health Teams provide nonprescriptive items such as deworming medicines, condoms and family planning pills to those who have been using them.

Information behavior related to coping with health problems

Women interviewees were asked whether they would prefer getting information or not if they were faced with a life-threatening disease or a serious health problem. All except two preferred to get information about causes, detection, treatment/management, prevention and coping with the illness or health problem. According to the monitoring and blunting theory, the information that might enable one person to deal with a problem (monitors) can be perceived as threatening by another person (blunters). The monitors therefore prefer getting information and they seek it, whereas the blunters generally do not or do so only to a limited extent (Baker, 1995).

Comments made by interviewees who preferred to receive high information input (monitors) were subdivided into actual and potential monitors. One interviewee living with HIV/AIDS was an example of an actual monitor. She pointed out that although she preferred and actually tried to get as much information as possible to help her to live and cope with the problem, she feared the public knowing that she had AIDS. Others pointed out that:

> For my ear problem, I try to get as much information as possible to enable me to make informed decisions. Right now, I am collecting information about doctors specialising in ENT so that I can have a choice. I have lived with this problem... and it is growing as I grow... getting information helps me to bear the problem and to become knowledgeable so that I don't make the situation worse... and if it is a terminal problem, I should know so that I get psychologically prepared
>
> **Woman 3, Masaka**

While the potential monitors pointed out that:

> Even if I won't survive, for example, if I had AIDS, I would try to get information that would enable me to prolong my life as the saying goes 'however short a time one can prolong life, it is better than death' (akaire mubulamu kakira emyaka mu magombe)
>
> **Woman 6, Iganga**

These comments agree with Baker's (1995) findings that "in uncontrollable situations such as multiple sclerosis, monitors will seek information whether they need it or not, because the information will give them an idea of how bad their MS may become. If and when the worst happens, they are prepared for it and the effects of the stress are less than they would have been if the monitors had not collected the information" (p. 73).

However, comments made by interviewees who preferred to get less or no information (blunters) were generally for potential situations, such as:

> I would not like to know much about the disease because the more I know, the more I would get worried... If the disease has no cure, nothing can change that fact; not even the amount of information one gets (woman6-lira); I can't go for an AIDS test, for example, because knowing that I am positive would make me very worried and die even earlier than if I hadn't known... so, I don't want to get information about a terminal condition since I am going to die... and it would be better if I am not told at all that I am suffering from such a disease. However, if the disease is not terminal, I would like to know how to manage it so that I don't make mistakes that might cost life
>
> **Woman 7, Masaka**

Information behavior: active and passive

Most information science literature discusses information needs in relation to active or purposive information seeking. However, the study on which the book is based revealed that although women had various health information needs, the tendency was to wait until information was accessed passively by coincidence, unless the need turned critical, then information would be actively sought. Information needs, therefore, do not necessarily always lead to active information seeking. This was also observed by Wilson (1997: 556), who pointed out that "whatever the situation in which a person perceives a need for information, engaging in information seeking behavior is not a necessary consequence... the individual's personality, perhaps coupled with other factors, may offer its own resistance to information seeking."

Furthermore, in a number of situations, the availability of information brought about the recognition and satisfaction of a previously unrecognized need. Hence, women reported to have accessed health information passively without first realizing that they needed it, but that after accessing the information they found it useful. For example, women go to church primarily to pray or to the trading centers for shopping, but on the way they see posters with information that later becomes useful. It was also noted that getting information repeatedly on the same topic from different sources tended to make people realize its importance.

> On my way to church, I came across posters indicating immunisation dates... After the church service, there were announcements about the immunisation dates and venue. When I went to a trading centre, I also saw similar posters indicating the immunisation dates... As I chatted with my neighbour, the immunisation issue also came up, but I had forgotten the exact dates... I went back to the poster and checked for the date
>
> **Woman 6, Lira**

This shows that although information was initially accessed passively and without realizing that it was needed, hence forgetting the details such as dates, later it became important and the interviewee became an active information seeker and went to get the immunization dates. This also highlights the dynamic nature of women's information behavior. Hence, the availability of information acted as a stimulus to active information seeking. This was confirmed by health workers who reported that many times, people ask them for advice about the information they had received from other sources.

In other situations, women found themselves receiving information (eg, from seminars) that they did not consider relevant to them personally at that particular time, but later it became useful. The findings have revealed that women's information needs were dynamic. They changed with time and situation, nature of illness/health problem, person involved (child, self and other family member), and from being latent to active or critical health information needs.

Consequently, information needs were of two major types: those that resulted mainly from critical incidents and triggered active information seeking, and those that the interviewees reported as unmet information needs but had not led to active information seeking. The former is referred to here as critical information needs, and the latter is regarded as latent information needs. However, approximately 27% of the interviewees habitually sought health information actively for general health knowledge and updating, whereas 14% actively sought information to cope with stress or a health problem. These and other needs may be simply referred to as active.

For this book, therefore, there was a need for a term or concept that would embrace passive access and active seeking or searching. The term "information acquisition behavior" would seem to embrace both modes of information behavior better than "information seeking behavior," which tends to focus on the purposive seeking/searching mode. However, not every information seeker acquires the information she/he seeks. Palsdottir et al. (2007) suggested that "information seeking can be broadly divided up in two information seeking styles: purposive information seeking ... and information encountering, which can be described as instances when people happen to come across information although they were not seeking it." That shows, however, that information encountering is not strictly seeking, but accessing information without seeking. Given this, this book adopted a broader or general term "information behavior," which encompasses the different styles.

Active to semi-active seeking behavior

A number of health needs that led to some active information seeking were highlighted by the women. For example, some asked health workers questions or read posters, whereas others talked with friends, relatives and/or neighbors about their needs. However, even after failing to get the information they needed, they did not engage in further seeking. Although such information needs remained unmet or unsatisfied, they did not lead to further active information seeking beyond the preliminary attempts.

Furthermore, it was noted that although women were predominantly responsible for providing health care within the family, their information behavior was greatly affected by the same families they care for and the resulting routines of everyday life, which is a paradox. For example:

> Caring for the family and attending to other domestic responsibilities sometimes tend to isolate me... I miss some meetings where information is disseminated
>
> **Woman 11, Bushenyi**

Some other information seeking acts were a result of failure to get satisfactory information or any information at all.

> I have consulted several health workers about my severe menstruation pains and failure to conceive but they just give me treatment without explaining what the cause is - and the treatment has not cured the problem... So, I need information about causes and where I can get effective treatment from (woman7-Masaka); I would like to get more information concerning false teeth... I have asked several health workers who keep telling me that false teeth don't exist in medical science, but I am not convinced because I have seen them... several children get them and they get treated by traditional methods and recover
>
> **Woman 11, Lira**

Other information acts resulted specifically from the problem of getting conflicting answers that left women's needs unmet and led to further information seeking.

> Some health workers tell us that it is necessary to give injections to children suffering from measles until the dose is over, others that once the measles 'come out' and show on the skin, you don't have to continue with injections... When I go to the clinic and find different health workers, they give me different advice about the same issue...I hope one day I will get a conclusive answer (woman1-Iganga); I asked during a seminar, whether hair dryers make mental illness worse, the seminar facilitator said 'yes', when I asked another one, she said 'no'; so I remained in doubt
>
> **Woman 4, Masaka**

Women's information seeking behavior usually included other players in the process; for example, consultations with relatives, friends and neighbors were common. Hence, interaction with social networks was an important factor in the information behavior of the interviewees. The information behavior of women, therefore, cannot easily be explained by single factors.

Active information seeking was highlighted further in the critical and active information needs subsection.

Passive access

Interviewees had information needs that did not automatically lead to active information seeking. Some responses to the question "what have you done to get the information you need" included:

> I would like to know why some women menstruate when breastfeeding, while others don't... I haven't done much yet apart from discussing with friends but they also don't know
>
> **Woman 9, Lira**

Other researchers have highlighted women's passive access to information in their findings. For example, Vogelsang and Oltersdorf (1995), while reporting the nutrition information needs of mothers in Germany, pointed out that even those consumers who perceived a need for information rarely became active information seekers. Information sources that required a high degree of active behavior on the part of the consumers did not reach many people because they preferred sources that were easily accessible and required only a little initiative from the individual.

2.3.3 CONSTRAINTS

The constraints category was defined in Section 2.2 and includes factors affecting access to health information. The factors were subdivided into negating and supportive factors. The negating factors were interpreted as the constraints, which were moderated by the enhancing factors presented in Section 2.3.4.

Negating factors

The factors that had hindered access to health information were identified as shown in Table 2.7b. Women suggested various ways in which the negating factors could be overcome or addressed, for

Table 2.7 Summary of Factors Affecting Rural Women's Access to Health Information

(a) Enhancing Factors	(b) Negating Factors
1. Presence of local council structure up to the village level	1. Economic factors
2. Personal attributes	2. Lack of time and heavy workload
3. Access to radio and its programs	3. Health workers' absence and behavior
4. Availability of seminars/workshops	4. Apathy
5. Educational factors	5. Social cultural factors
6. Membership to active groups/NGOs	6. General unavailability of information
7. Religious leaders, beliefs and practices	7. Character, emotions and attitude
8. Spatial/geographical factors	8. Geographical factors
9. Interpersonal factors	9. Late invitations to workshops
10. Social factors	10. Educational factors
11. Active women leaders	11. Language barrier
12. Economic factors	
13. Health education at regular intervals	
14. Mobile phones/new technology	
15. Access to Government programs	

example: making health services more accessible to the rural communities; the need to train and deploy more community health workers; the need to improve household incomes at the national level and the provision of simplified printed information in vernacular.

Supportive factors

The factors, summarized in Table 2.7a were identified by women as those that had enhanced their access to health information and, hence, acted as moderators. The factors are listed in Table 2.7a in order of importance, with the first having been mentioned the highest number of times. The enhancing factors are initially outlined in this section to ease comparison between them and the negating factors, but they are finally presented in Section 2.3.4. Women also made suggestions on how such factors could be strengthened, for example: capacity building by training LC officials, religious leaders and women leaders in information collection and dissemination; more sensitization of women to participate actively in activities where information is disseminated; more mobilization of women to participate in adult literacy programs and more sensitization of spouses to support women's information activities.

As providers of information, health workers were asked to report, from their experiences, what they considered to be factors affecting access to health information by rural women. Most of the enhancing factors identified by the women (Table 2.7a) were similar to those identified by health workers. The major difference was in the emphasis; for example, fifteen women highlighted access to radio as an enhancing factor, whereas only one health worker did so. Eight women identified accessibility/spatial factors, but only one health worker did. Furthermore, factors such as interpersonal, social, and availability of seminars/workshops and other training opportunities were only identified by women who had experienced their enhancing role.

Similarly, there was common ground among the factors hindering rural women's access to health information, which were identified by women and those identified by health workers. However, health workers did not report leadership and late invitations to seminar/workshops as negating factors. This was not surprising because women, rather than health workers, had been the victims of such problems. Furthermore, language barrier was reported by women as a distinct factor, whereas health workers generally reported it as an educational problem.

Further analysis of data indicated that there were two types of constraints, namely: constraints to information access and constraints to information use. They are summarized in Fig. 2.2.

As Fig. 2.2 shows, *constraints to information access* were subdivided into the following:

i. Constraints concerning specific information sources
ii. Socioeconomic constraints

Constraints concerning specific information sources

The most commonly identified constraints were related to health workers, printed and audio-visual sources, seminars and libraries, in that order. For example, women reported that health workers' absence in rural areas constrained their access to information. This was in reference to professional health workers rather than the TBAs and the village health team members who were readily available in the villages. The absence of professional health workers meant that women had to go to health units that were, in most cases, located far away. Some health workers confirmed that issue

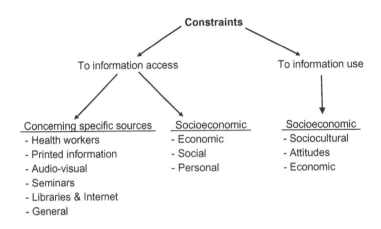

FIGURE 2.2

Analytical representation of the constraints to information access and use by women.

as a challenge and highlighted heavy workloads, understaffing at health units, transport problems and others, as the following comment shows:

> We find it difficult to regularly go to rural areas for health education because of the heavy workload and understaffing at health units, transport to some areas is a challenge, but we try our best although we are not able to satisfy the demand
>
> **Health worker 5, Masaka**

With regard to printed sources, the major issues were unavailability and expense of books and magazines. Posters, however, were free of charge but unavailable to a number of interviewees who reported that they only saw them when they went to health units, and seminars, as illustrated in the following comment:

> There is no printed information such as magazines, pamphlets or books but we see posters at the church. Sometimes when I tell people issues concerning health, they ask me 'how did you know about it'? If I had some source e.g. a magazine or booklet, I would just open the page and show them, but without an authentic source, it is difficult to disseminate information. People want to see the source to believe you. Sometimes I show them my seminar notes and they get convinced about what I tell them ... On this village there are 52 adult women, and only 10 can't read or write, so the rest of us can benefit from printed information if it was available, and it would make our work as community leaders and mobilisers a lot easier; in any case, we who are literate would read and pass on the messages to others
>
> **Woman 10, Masaka**

These findings agree with those of Carter (2002), who reported that "when groups with no literate members were asked if printed information was seen as important, they stressed their desire to obtain materials: 'The school teacher or our children can read for us'... Print is a well trusted source of information ... it is a relatively permanent method of sharing information."

To the rural people, therefore, the issue was availability of printed information rather than illiteracy. Furthermore, expense and lack of accessibility of printed sources were problems, as already indicated.

Similarly, several constraints were identified regarding audio-visual sources such as CDs/DVDs, films and television. Some women pointed out that although there were many audio tapes with health messages in the vernacular recorded as music and plays, they rarely listened to them; usually, it was their husbands or sons who used them to entertain guests. Others pointed out that they could not afford them.

Although radio was ranked highly as a source of information, some interviewees from Lira district raised the issue of limited health programs in the Luo language on radio that reduced their access to information from that channel. Six other women from Lira and Iganga reported that they rarely listened to the radio for various reasons. That showed the difference between interviewees from Bushenyi and Masaka, who all reported that they listened to the radio regularly. The other constraint with radio, as indicated under information source preference, was the lack of interaction or feedback. Even when there were phone-in sessions, interviewees complained about telephone lines being busy all the time, making it difficult for them to "talk back" to the radio and ask questions. Such issues attracted media audience studies, as cited by McQuail (1994), who reiterated that "the growth of audience participatory programming by way of phone-ins and studio audiences is evidence of the favour in which the appearance of audience contact and interaction is held. There has long been talk of even greater possibilities of audience feedback, based on interactive telecommunications" (p. 324).

The constraint concerning libraries was unavailability, as highlighted in Section 2.3.1. There were no libraries in the eight rural subcounties studied, and several interviewees reported that "*it is difficult to get information in books as there is no library nearby.*" However, unlike Bushenyi and Iganga, there was a public library in Lira town, although half of the women interviewed in Lira did not know about it, four knew about it but had not used it, and only two had used the library (Woman 3 and Woman 11, Lira). It was also noted that there was an NGO resource center 4 km from Lira town that had been used by one of the interviewees. Similarly, Masaka had a municipal resource center in Masaka town, but only the interviewees from a subcounty nearby knew about it; the rest (50%) did not know about it. Two women who were primary school teachers reported using their school libraries/book collections, and two others reported knowing about the municipal resource center although they had never used it. Hence, absence of libraries in rural areas, distance between the rural areas and the public libraries in towns, and low education levels combined to negate libraries as a source of health information among the interviewees.

The women also pointed out that the Internet was a potential source of information due to cost and related factors, as indicated in the section about economic and technological constraints.

The other general constraints were mainly about unavailability and inaccessibility of information sources and/or lack of knowledge about the sources in general. They relate to comments made in Section 2.3.2 (Information Needs) about the most difficult type of information to access. Furthermore, some women raised the issue of selective information seeking as a problem, for example:

> I tend to select only the information I want to use immediately, and leave out the rest... the disadvantage is that later when I realise that I need the information I left out previously, I don't always get that information. For example, recently, I ignored a brochure on short sightedness, now a friend has just asked me whether I have any information on this topic... When I went where I saw the brochure, they were finished
>
> **Woman 5, Lira**

Socioeconomic constraints to information access

Constraints to information access, other than those concerning specific information sources, were broadly classified as economic, social and personal constraints. Analysis revealed some overlap and interdependencies between constraints, for example, personal and social (mainly gender and culture). There was also an overlap between constraints concerning specific information sources and socioeconomic constraints to information access, such as economic constraints that hindered women from attending seminars/workshops. The most constraining factors, as Fig. 2.2 shows, were economic, followed by social and personal issues. They are highlighted in that order.

Economic constraints included mainly financial limitations because they affected access to seminars/workshops, mass media and consultation. For example, one needs money for credit to be able to consult a health worker by phone or for transport to a health unit. Lack of money to attend seminars that were not fully funded was also a problem:

> I was invited to attend a seminar at Bukulula, about 20 miles from here, but I didn't have money for transport; so, I missed... Seminars which are nearby are easier because one can just walk there
>
> **Woman 1, Masaka**

Lack of funds, therefore, constrained access to specific information sources. If financial hardships could be mitigated through sustained economic growth, other things being equal, a positive chain reaction would be created to enhance information availability, accessibility and use.

Two other economic-related constraints were spatial/geographical and communication/technological, which were generally a result of the country's economic situation. For example, distance from the district head office greatly affected access to information. The findings showed that women in subcounties nearer to the district head office generally had easier access to health units, libraries where they existed, and posters, whereas those from subcounties more than 15 km away from the district head office identified printed information as the most difficult type of information to access. Many reported that they hardly ever received any printed health information, whether they were books, posters or newspapers. Distance affected attendance of seminars/workshops as already indicated. It also affected transport to and from rural areas, which made it difficult for the women to reach information sources and the health workers to reach remote rural areas. Spatial factors were compounded by the poor road network/transport and limited communication/technological facilities in some rural areas where, for example, the telephone network signal was poor.

> This place is not easily accessible; hence public transport is scarce and expensive. In the rainy season, there are usually floods, which make the situation worse
>
> **Woman 1, Bushenyi**

Social constraints included time and heavy workload, gender/culture/religion, and educational and language factors in the order of importance, as Table 2.7b shows. Women pointed out that the amount of work and responsibilities they had greatly limited the free time at their disposal. Consequently, it was difficult for some women to regularly attend meetings, seminars or workshops, watch films or listen to radio. For example:

> Sometimes I switch on the radio and even fail to pay attention when I am attending to children or my customers in the shop... I have also missed seminars twice this year because of my parental responsibilities: the first one, I was attending to my son who was admitted in hospital just before the seminar; the second time, my daughter was sick here at home... I couldn't go... when I get help from my family members, then I attend
>
> **Woman 11, Masaka**

Apart from economic factors, the other constraints concerning seminar/workshop attendance included the multiple responsibilities of women and appropriate timing such that during planting and harvesting seasons when rural women would be busy, seminars are not scheduled on weekdays. The timing of seminars was closely linked to a time and heavy workload constraint. Furthermore, a combination of professional (eg, primary school teacher), domestic and community work also burdened women and denied them time to access health information:

> During the week I have to prepare my school work and to do some marking now and then; at home, I have to do the domestic chores. By weekend when some seminars are held, I am tired... actually my weekends are dear and I find it difficult to attend some important seminars or to do my LC work... It is a real sacrifice to the community but sometimes I fail to find time to sacrifice!
>
> **Woman 5, Iganga**

Poor time management was another problem. Although it was noted that time limitations resulting from heavy workloads constrained women from accessing information, some women pointed out that the problem was compounded by lack of time management skills that would enable women to optimally allocate activities between time-competing needs. It was recommended that:

> Women need to be sensitised about time management and finding a way to reduce the domestic work load, e.g. by using time saving devices or appropriate technologies that are affordable... this will reduce the burden of domestic work and enable women to get some free time which would benefit the whole family in the long run
>
> **Woman 1, Lira**

Gender and culture created impediments, as did, to a less extent, religious practices. It was pointed out that some husbands discouraged women from attending meetings or seminars. Consequently, the women councils in different areas reported having embarked on a sensitization drive that was gradually improving the situation. Furthermore, women, particularly the nonprofessionals, shared some challenges concerning their socialization. They pointed out that men benefit from going out where various issues are discussed and information disseminated. The cultural norms and practices do not support or encourage ordinary women to freely go out, which affects their information activities, as indicated in the following comments:

> the way we were socialised stop us from getting information - as girls, we were brought up in a protected way, not to go to places where we could be tempted.... When we grow up, we tend not to be outgoing... the position in which society places women - we receive and don't go out to seek, unlike men! Even when we go out for meetings, for example, some of us keep quiet, we don't ask questions. Furthermore, polygamous marriages, in which we find ourselves, bring about rivalries and force us to succumb to some negative practices which reduce our opportunities to access health information... we need to change though, but will society accept, shall we not get even more discriminated!
>
> **Woman 12, Lira**

Indeed as observed by Conte (2015), Ugandan women find themselves at a crossroads; with dignity, they play a complex social role, highlighting the values of their upbringing but with stories about a rapidly changing society in which they serve both as guardians of culture and harbingers of reform.

Religious practices were reported to have denied some interviewees access to information, for example:

> The church I go to discriminates against unmarried couples; so, some of us never get a chance to attend seminars organised by Mothers Union and other church seminars because we are not invited
>
> **Woman 2, Iganga**

With regard to educational constraint, it was noted that eight women from Iganga, six from Bushenyi, five from Lira and four from Masaka were primary level leaders, one from Iganga had no formal education but had attended functional literacy classes, and only eight were professionals (three from Bushenyi, two each from Lira and Masaka and one from Iganga). Low educational level tends to lead to low self-esteem, which could demotivate the individual such that she does not seek information actively. Some women pointed out that:

> Low literacy levels affect the amount of information one gets, say, from a seminar because one is not able, for example, to write down notes for future reference; some women cannot read either in vernacular or English or both, so they can't follow prescriptions or read drug leaflets... they miss information in printed sources
>
> **Woman 7, Bushenyi**

Though related, women reported language and education as distinct factors that constrained their access to information. This was because information disseminated in local or international languages, which people did not understand, was not accessed. Sometimes the interpreters do not do a good job, making it difficult for people to understand, as the following comment shows:

> Recently, we went to watch a video at the health centre which was in Runyankole – and yet we speak Lusoga in this area, it was not easy to follow... Some seminars or health talks are conducted in English: last month, a talk about FP was organised here, but the presenter couldn't speak Lusoga or Luganda; so we complained and one of the LC officials started translating but he admitted that he couldn't translate some things... at least if it was a health worker translating it would be better, so, one gains little from such seminars
>
> **Woman 7, Iganga**

Furthermore, language was reported to have stopped some health workers from delivering information to the communities where they worked, for example:

> Although I work in this area, I don't come from this region; so, I do not speak the language... I have not learned the local language yet. This stops me from holding health education sessions in rural communities until I get an interpreter
>
> **Health worker 9, Bushenyi**

Limited appropriate health information materials in local languages were also identified as an issue. It highlights one of the consequences of lack of a national language, which could be used to disseminate information to the whole nation. Information providers, such as radio stations and the MoH, have had to go to great lengths to try to reach as many people as possible. Some pamphlets and posters on AIDS, cholera, FP, malaria and immunization had been translated in the different languages and some women and health workers reported using them. However, the numbers produced did not seem to reach everybody. The multitude of languages in Uganda continued to pose serious challenges in translating health content (WHO, 2006).

With regard to personal constraints, there was an overlap between personal and social factors that constrained information access as already indicated. The major issues were apathy and character. Some interviewees referred to it as apathy, but if apathy[7] means "absence of interest or concern; indifference," then apathy probably was a symptom, and what causes it is the lack of awareness of the information sources. Although Sorrentino et al. (1990) pointed out that "... many people are

---

[7]Apathy is usually acquired, whereas passivity is more of an inborn character; for example, lack of awareness of the importance of information leads to apathy (not passivity) because if one became aware, then the situation would change.

simply not interested in finding out information about themselves or the world...," some rural women in Uganda seemed not to be aware of some information sources; many were preoccupied with other more pressing demands on their time; therefore, it would not be appropriate to describe them as "simply not interested." The issues at hand were as already stated (namely, poor time management and genuine lack of time) as well as the lack of awareness of some information sources.

Women also reported that their emotions, attitudes, character and perceptions sometimes constrained their access to information. Examples of comments about attitude and emotion follow:

> if we, as rural women, are to improve our access to health information, we need to be vigilant and stop feeling inferior, incapable or incompetent
>
> **Woman 5, Lira**

> stigma... but also fear of the public to know that I have 'slim'[8]... Although I know that I could get valuable information from TASO, I have not gone there yet; I fear meeting people I know there... this actually stops me from asking openly... I only talk to health workers and in confidence
>
> **Woman 4, Bushenyi**

During the analysis of the interview data, it was difficult to judge whether some of the attributes or characteristics reported were permanent, common or regular enough to be classified as one's personality. Hence, the constraint was women's character.

Low or negative self-concept/perception was another constraint. This was referred to as self-concept because it was what the women felt and reported about themselves, and it was not possible to confirm or refute it during or after the interviews. What came out clearly during the interviews was that women needed more confidence and self-esteem to access information. The perceptions included low self-esteem, lack of confidence and self-pity, as the following comments show:

> I am quite shy ... and sometimes I feel that I have nothing constructive to contribute so I rarely ask questions or make comments during meetings
>
> **Woman 4, Lira**

> maybe I pity myself too much as a lame person, sometimes I just decide not to go for meetings because of the difficulties I have in walking... but probably I could initiate something for my disability post on the LC rather than waiting for programmes initiated at higher levels
>
> **Woman 2, Iganga**

Constraints to information use

The fact that sources of information are available and accessible and information is processed (incorporated into the user's framework of knowledge, beliefs or values) is no guarantee that the information will be used and lead to changes in the user's state of knowledge, behavior, values or beliefs, as observed by Wilson (1997). It was noted that although some information was reported to have led to changes in the user's state of knowledge, a number of constraints intervened to stop some users from putting the knowledge into practice. The most important constraints to information use were sociocultural, followed by personal factors, such as attitudes and perceptions, and economic issues.

---

[8]HIV/AIDS is commonly known as "slim" in Uganda.

Sociocultural constraints to information use

These were mainly gender issues and, to a less extent, other social constraints.

The main gender concerns were about FP, hygiene, STIs and condom use. In most cases, husbands stopped women from putting their knowledge into practice or they did not cooperate. In a few situations, such as FP, however, women's own values made them feel insecure about stopping child birth, therefore constraining information use.

> The information I got from the FP clinic made me aware of the dangers of giving birth to too many children, and that it is important to space even the few children one plans to get; however, I have not been able to put that knowledge into practice because my husband insists that he still wants children and he has stopped me from going back to the FP clinic and from using the pills I got; yet my current interest in FP is for spacing, rather than stopping births, but he doesn't understand and he is not even willing to
>
> **Woman 8, Lira**

> Although I would prefer a permanent birth control method, because I have 5 children already, I fear to stop giving birth completely because my husband might get another woman and start producing
>
> **Woman 12, Bushenyi**

Some health workers highlighted similar gender constraints and pointed out that there was a need to promote equity in participation in FP services and sensitize both men and women to participate in the decisions concerning their families, for example:

> a certain moslem man had three wives whom he encouraged to take FP but they refused because they were competing to have more children than each other... so the man secretly decided to have vasectomy
>
> **Health worker 3, Bushenyi**

Furthermore, treatment or control of sexually transmitted infections was a problem, as well as condom use. Some women experienced resistance when they tried to convince their spouses to go for STI tests or use condoms. Male resistance to condom use in Uganda had been reported by a number of authors, such as Standing and Kisekka (1989), Kaleeba et al. (1991) and Marcus (1993). The situation gradually improved with the successful national HIV/AIDS campaign.

Social constraints raised issues of self-control, for example, to be faithful to one partner. Related to social constraints was religion. It was noted that religious practices and values did not come out explicitly in the women's data as constraints to information use, but they were reported by one woman as constraints to information access and by several women as moderators of information access. However, some health workers sounded frustrated about the conflicting messages people continued to get from the church and health workers that, in their view, constrain information use as the following comments show:

> What we tell people to do contradict the stand of the church which emphasises morals and disapproves particularly condom use and FP which is detrimental to life... This affects the use of contraceptives and puts women's lives in danger because of frequent child births... When we go out health educating and encouraging people to use condoms or FP, they ask us questions which make us feel like we are not followers of the same church!
>
> **Health worker 2, Masaka**

It was noted that people's attitudes stopped them from using the information accessed. At a more personal level, the attitude or views held by interviewees about some health programs such as immunization included "tokamanya" in Runyankole (vernacular in Bushenyi district), which means "you never know"; this referred to one not being able to predict what will happen after the immunization. Information about polio immunization was accessed by all the women interviewed. However, there were still unanswered questions about the health benefits of immunization, but more so about the safety of the vaccine; for example:

> There are many rumours that the vaccine is not safe... it was contaminated with the AIDS virus or other dangerous chemicals...
>
> **Woman 9, Masaka**

> Although I have heard that immunisation protects children from polio, etc., I am not sure what it actually does because my first born who was immunised still got polio — she limps; I have no explanation for that!
>
> **Woman 1, Bushenyi**

Some women reported that they needed information about the safety of the vaccine, as indicated in Section 2.3.2. Such views affected the use of information to the extent that some people did not take their children for immunization even though they had known about the immunization program that year. These were unmet information needs that made people develop attitudes and views that constrained information use.

With regard to economic constraints to information use, some women pointed out that, due to limited financial resources, they were not able to use some of the information they had accessed or to apply the knowledge they had acquired. For example:

> In one of the seminars I have attended, I learned how to extract milk from soya beans which is an alternative way of getting milk in a rural setting especially during the dry season when cow's milk becomes difficult to get. However, soya is scarce in this area ... one would have to buy it, which is expensive for me. So, I only use this method occasionally when I get cheap soya
>
> **Woman 1, Lira**

2.3.4 MODERATORS

As defined in Section 2.2, the moderators acted as a buffer to regulate, reduce or intercept the constraints to information access and information use. The analysis revealed a relationship between the constraints to information access and the moderation by individuals, organizations and structures that reduced or intercepted the constraints and led to improved information access. For example, the problems of limited access to information caused by having few health workers reaching some rural areas, lack of time for the women to attend meetings/seminars and watch audio-visual sources, language barriers to information provided in foreign languages, and others led the LC executive committee members to take on an information dissemination role (for the benefit of their communities) either by inviting health workers to give talks in the LC meetings or by the LC executive members moving door-to-door to ensure that information reaches every member of the community. At a slightly higher level of abstraction, such a relationship seemed to be one in which the value of information, the need for information access and use, and the prevailing constraints in rural Uganda had led, among other things, to the institution and flourishing of an informal mechanism of health information provision.

Were (2014) highlighted the importance of creating new community norms in Kenya that promote health and hygiene and prevent disease, and that it was possible to achieve that by empowering communities contrary to most professional opinion. Nuijten (1992) had observed that such local practices or initiatives are often denied their due importance and labeled as disorganized, traditional or indigenous in development studies literature. Such debates, however, remain far removed from the everyday practice of the people, as this book has demonstrated.

Moderators were generally identified as enhancing factors in Table 2.7a and are divided into two:

i. Moderators of constraints to information access
ii. Moderators of constraints to information use

Moderators of constraints to information access

They were subdivided into the following groups: leaders (local, women and religious); personal and interpersonal attributes; rural outreach programs; educational; geographical; social; technological and economic moderators. There were also interdependencies between factors and collaboration between different moderators.

Leaders as moderators

Data analysis showed that leaders were the most important moderators. They included LC, women and religious leaders.

Referring to the LC structure, both the women and health workers identified the supportive Government policies and the resultant leaders as moderators. They pointed out that Ugandan's Government support to women and the affirmative action put in place, for example, the 30% representation by women on all the LCs from village to parliament, had given women an opportunity to participate in local and national politics and in decision making. Furthermore, the general public sensitization regarding women's role in Ugandan society had reportedly encouraged some men to allow their wives to participate actively in community and national leadership activities, which gave them an opportunity to access different types of information including health. At the time of data collection, it was noted that in the Bushenyi, Masaka and Lira districts, for example, the LC5 secretary for social services, who was the district head of social services under which health falls, was a woman. Another example was the health unit management committee at the health unit level, which had at least one woman who represents women's interests and also collects and provides information to and from the women she represents.

The LC officials, for example, the secretary for women's duties and responsibilities, are responsible for mobilizing and sensitizing women primarily and other members of the community regarding different aspects of life, whereas the secretary for rehabilitation was responsible for health in the LC structure. Furthermore, the secretary for information as well as the other executive committee members were responsible for, among other things, mobilizing, informing and maintaining good health and welfare of the communities they served. They held meetings, organized house-to-house visits, drama and film shows. Women also reported that the vigilance of LCs made a lot of difference in the provision of health information to rural areas. This was confirmed by health workers who pointed out that:

> whenever the LCs identify a problem, they inform us... In fact they are the ones who requested for the current outreach activities we are carrying out... When I delayed, they even threatened to report to the DMO; so, these rural outreaches are demand — driven
>
> **Health worker 4, Iganga**

Active women's groups or clubs and NGOs were reported to organize meetings, talks and seminars/workshops about different aspects of health. Some women's groups also had drama activities and staged plays about various health topics, and some presented up to the national level. They pointed out that:

> At the village, the shows are free because we are rehearsing and also awareness raising, but outside we charge reasonably... Women's turn up is very good: about 70% probably because it is a women's group involved. We, members, learn a lot about the topic we are focussing on because the script is written or edited by health workers
>
> **Woman 7, Iganga**

With a few exceptions (namely the issues raised by one health worker under constraints), the role played by religious institutions in moderating information access was greatly commended by the women and health workers interviewed. This was in relation to both the direct provision of information and to the shaping of beliefs, attitudes and behavioral change that led to health promotion. For example, health workers reported that:

> While condemning superstition, the church encourages people to seek advice or treatment from health workers whenever they or members of their families fall sick; hence, religious leaders are a key in mobilising the masses to use health facilities and in changing or influencing their health seeking behaviour
>
> **Health worker 1, Masaka**

Women pointed out that religious leaders moderated information access by preaching, organizing drama or film shows, supporting community religious groups and visits, and collaborating with health workers, LCs and Government. The majority reported church leaders and the few Moslems who were interviewed reported leaders in their faith as the moderators.

> They always preach about current health problems such AIDS, nutritional deficiencies and cholera, and highlight ways these problems could be prevented; for example, adultery could lead to AIDS and other STIs... The message is very clear; so, one can't miss it
>
> **Woman 4, Lira**

> they organise regular films and drama shows on different health topics, and these are free of charge, which makes this channel accessible to both the rich and the poor
>
> **Woman 8, Bushenyi**

> they support and inform us through women's groups e.g. the Catholic women's guild... Saint Matia Mulumba project provided 3 bore holes on this village and talks about safe water; while Saint Kizito provided seminars on nutrition, etc
>
> **Woman 4, Masaka and Woman 10, Iganga**

In addition to religious leaders, it was noted that religious beliefs and practices played a role as social moderators. Ugandans are generally strong believers, especially women, who regularly attend religious functions, consult religious leaders on spiritual or welfare issues including health, and participate in various religious activities. Religious functions were identified as the second-easiest way women accessed information (in Section 2.3.1). The presence of the church and its extension agents up to the grassroots level tended to sustain the practices and nourish women's beliefs, which facilitated information access from religious sources.

Collaboration or cross-fertilization (to use the exact word used by a primary school science teacher) between LCs, health workers and the church greatly enhanced health information dissemination. For example, announcements were made after a service by health workers or LCs, and the church was acknowledged for providing space for meetings and various other support.

Furthermore, in the subcounty where some LCs had been considered inactive, church leaders were commended for performing sensitization on various health issues, organizing training sessions, announcing health events including immunization days and venue, and organizing other activities relevant to health information delivery.

These points were confirmed by health workers who reported that they had witnessed the preaching about various health issues in church, and that religious leaders facilitated their work greatly. For example, to avoid going from house to house, messages were given to the church or mosque; they provided free venues for health workers to hold health education sessions, they organized sessions, and they invited the people.

> Actually we, or even the LCs, get a lot of assistance from the church... Several messages are prepared and given to the church leaders to announce because many people go to church... it saves us from going door to door.... Later when we ask people how they knew about certain issues, they tell us that they heard from the church directly or somebody who was in church informed them
>
> **Health worker 4, Iganga**

Personal and interpersonal moderators

Personal moderators included attributes such as being active, practicing what was learned, and openness or interest in sharing. Approximately 15% of the women interviewed pointed out that their active, outgoing and/or talkative nature moderated their access to information. Some gave the example of being elected on LCs or other leadership positions and pointed out that these gave them opportunities to attend seminars and other meetings where information was disseminated. To be elected requires one to have been active in the community, they added.

> Being active in clubs, LCs and other community activities is an asset... I am quite inquisitive and mobile ... personal interest, involvement and enthusiasm provides me with many chances to access information
>
> **Woman 11, Masaka**

Practicing or implementing what was learned also moderated information access and motivated women to get more information. Some women pointed out that when they learned something, they tried to put it in practice, and in the process they became more knowledgeable. Furthermore, in implementing what one learned, it opened various other opportunities, such as:

> when seminar organisers or sponsors come to check and find that one has done something as a follow up to the seminar or course, they invite you for other seminars, which brings in more information
>
> **Woman 1, Masaka**

Openness to new ideas or interest in sharing and learning enhanced access to information. It was also noted that personal attributes moderated some negative characteristics, which changed through experience and enhanced information access as indicated:

> Attending seminars regularly and participating in various community activities has given me confidence... I can now stand in public and talk confidently; I am no longer shy ... This gives

> me many opportunities to be selected on committees where I attend more meetings, seminars and interact with more people and learn more
>
> **Woman 12, Masaka**

Interaction with other people also moderated information access. For example, relatives, friends and neighbors were commended for sharing information about what they heard on radio, watched on TV and learned in meetings, and for their willingness to be consulted or asked questions.

Rural outreach moderators

These included radio and national events. To increase coverage, radio companies introduced upcountry "FM" stations that moderated access to information. Such stations were Voice of Toro (in Western Uganda), Gulu FM (Northern Uganda) and Central Broadcasting Service (Central Uganda), in addition to the national station, Radio Uganda, which was accessed throughout the country. The upcountry FM stations addressed the language problem because they broadcast in the main local language of the area where they are located. Many women emphasized the importance of listening to radio, as already indicated, and added that the presence of several radio stations gives them an opportunity to select programs of interest to their needs. Health workers also pointed out that radio had moderated people's access to information and supported the health education sessions as the comments show:

> When I hold health education sessions, I notice that attendees who listen to radio regularly contribute information quoting the radio, which enriches the discussion. This makes us encourage other people to listen to such programmes to access the information... others come to us and seek more information or advice about what they heard on radio; so, it becomes easier for us to reinforce what they already know, and it makes them seek for information
>
> **Health worker 2, Bushenyi**

There were also programs and national events held in rural areas, which included Government projects and programs, such as those focusing on AIDS and other STIs, water and sanitation, and functional literacy. In addition, there were national health-related events and community service days organized periodically. Women's participation in such activities moderated their access to information, which they would not otherwise access.

Health workers also pointed out that women's participation in such events was important in moderating access to information on specific topics. They cited the annual World Health Day, World AIDS Day, World Population Day and others, which enabled women and the general public to learn about the health issues on which they focused.

Educational moderators

Literacy and formal education, school children, seminars/workshops and health education in clinics all moderated information access in various ways. Oxaal and Baden (1996: 21) pointed out that "women's social status, self-image and decision making powers may all be increased through education, which may be key in reducing their risk of maternal death, resulting from early marriage and pregnancy or lack of information about health services."

Literacy and formal education was highest in Masaka district, where 75% of the interviewees identified it as an enhancing factor to information access, in comparison to 17% of the interviewees from each of the other research districts. This generally agrees with the 2014 Uganda National Census report, in which Masaka district average is higher than that of the rest of the research districts.

Women pointed out that education and literacy moderated their access to information greatly. They were able to access information in printed sources and record seminar proceedings for future reference, and the professionals reported the advantages of their professional training that moderated some of the constraints that had been reported by other women, such as:

> Ability to read, write and understand helps me a lot in collecting information for myself and my community wherever I go... in accessing printed information from books, magazines, etc... even when I miss seminars, my friends send me their notes or handouts to read and I write some of them for future reference
>
> **Woman 5, Masaka**

Others had benefited from functional literacy programs that enabled them to read and write. School children also act as moderators to reduce the illiteracy problems among parents or family members. They moderated information access by providing information through concerts staged at schools or in other places, and sometimes on radio or TV. Their literacy skills and school books also moderate information access as the children help to read or translate for parents who are not able to or share what they learn at school.

Seminars/workshops have become an important educational medium and one of the major sources of information. Women (more than 50%) pointed out that the availability and accessibility of seminars was a moderator; however, the need to hold such training opportunities at lower levels, such as parishes, so that many more people benefit was expressed by interviewees in subcounties far from the district head office. Furthermore, the importance of the health information training sessions conducted by Makerere University Library and the College of Health Sciences could not be emphasized enough by the women who attended. They pointed out that the sessions greatly supported the work of LCs, which in the end promoted and improved health in the community. The fact that the sessions were free of charge moderated the financial constraints and made it possible for women to attend. Such sessions were recommended to continue to update the knowledge of community leaders given that the people involved are adults who need repeated or regular updating.

The existence of training programs for rural people was also identified as a moderator by health workers. They pointed out that the capacity building projects, for example, that trained the community (LCs and other civic leaders from parish to upper levels) and equipped them with skills to collect, record and use information, such as on different health and development issues, were very important because they helped to improve the information management skills and to sensitize community leaders about the importance of information. This enhanced information access by rural people.

Furthermore, health education combined with regular clinics for antenatal, FP and immunization conducted in health units was commended for providing women with the information they may not otherwise have had a chance to access. The importance of such information made the women demand health education from health workers who conduct rural outreach programs, thus making such demands active information needs.

Spatial/geographical moderators

It was highlighted under constraints that people living nearer to the district head office were able to access various types of information. Short distances to the clinic, hospital and other health units made it easy for people to reach these sources of information; therefore, the distance moderated access to information.

Social moderators

While promoting the cohesion of society, social factors moderated information access in a number of ways. Several women pointed out that social gatherings and functions such as weddings or burials were some of the easiest ways they accessed information. This was confirmed by several health workers who reported that those functions are targeted by information providers because they know that many women attend them. Other women highlighted the role of their families and religious beliefs and practices as moderators. For example, support from the family enabled women to spare time to attend seminars/workshops where they accessed information. The women who did not get such family support had raised it as a constraint.

Technological moderators

The main technological moderator identified by women was the mobile phone, as indicated in Section 2.3.1. Although women had not directly benefitted from the vast amount of health information available on the Internet, they pointed out the difference brought about by the mobile phone, and health workers did the same. The presence of a mobile phone facility in almost all households in the areas studied was a positive technological development that had greatly increased the women's opportunities of accessing information as well as their ability to communicate, consult, share and disseminate health information to the community and higher levels in a timely manner, as the following comments show:

> I now access the radio on the mobile phone, so I can listen to a programme of choice unlike in 2000 when a household had one radio and listened to programmes selected by the head of the household which were mainly announcements... old people like announcements very much
>
> **Woman, Iganga**

> we can now call a health professional for advice on what to do with a patient who is not improving after treatment, where to take a patient with a rare condition or problem, etc without wasting time and this saves lives
>
> **Woman, Bushenyi**

> I send information on phone and receive a response on phone, without bothering to move, no transport, no time wasted... this is great... for me the major change is that I have a mobile phone and an email address
>
> **Woman, Masaka**

> Women sometimes fail to raise money for transport to attend FP clinics. Mobile phones have greatly eased communication; for example, some women bleed after starting or changing FP method yet their husbands do not allow them to frequent towns where the clinics are, in fact many go on FP secretly because their husbands are not supportive, that is why the injection method is preferred by many. Such women now ring the midwife/ health worker and ask for advice... Poor transport and lack of FP clinics in some rural areas had hindered the progress of FP, but the mobile phone has made a real difference
>
> **Doctor, Masaka**

At a country level, the introduction of the Mtrack electronic system by the MoH was also identified as having greatly eased communication between the MoH and lower levels in the community.

While at the district level, various health messages were sent by the district health information system (8008) on community leaders' mobile phones about various public health issues, for example, announcing outbreak of measles, cholera and Ebola in the neighboring Congo and other epidemics, forthcoming events and others. The facility increased access to accurate information from the MoH through the district and greatly reduced the isolation experienced previously.

Economic moderators

As reported in Section 2.3.3, many health information access factors revolved around money, for example, buying batteries for the radio, credit for the mobile phone, paying for video or drama shows, consulting health workers and transport to attend seminars. Employment and/or income-generating activities such as projects, revolving loan scheme, trading and crafts enabled women to improve their incomes and to access health services and information, for example:

> With the little money I get from my simple projects, I am able to save and buy batteries for the radio... I do this myself because I enjoy listening to radio ... batteries are cheaper than newspapers which have to be bought daily yet with batteries, I can go on for about two weeks (woman3-Masaka); The interest free loans and the revolving loan scheme provided by our club/group have assisted us a lot in starting small businesses... try to improve our well-being and that of our families
>
> **Woman 1, Bushenyi**

Finally, these findings have revealed some close collaboration between individuals or interdependencies between various moderators, which by themselves enhanced information access directly or indirectly by reducing the constraints. Furthermore, the provision of information in different formats to cater to the different capabilities and interests of people moderated its accessibility. This was well demonstrated by information about immunization and AIDS, which had been accessed from various sources by all the interviewees.

Moderators of constraints to information use

The major moderator of constraints to information use was the value of information itself. This made people moderate information access and, in the process, they moderated information use. For example, when information providers translated and simplified the information they provided, such as during health education talks, seminars, drama, and what is preached in church, they repackaged information and put it in a form that was not only accessible but also usable by the women. The quote that follows from a woman in the Bushenyi district about the LCs is another good example. Furthermore, it was reported that the MoH provided information that had been simplified and translated in the vernacular so that women could understand it and be able to use it for composing songs or plays to disseminate the information further. Hence, besides its value, the quality of information and supportive infrastructure moderated information use. A few examples are highlighted here.

Quality of information

The quality of information received assisted in overcoming misconceptions or negative views, which led to information use. Clear and complete information that was easy to understand and the full explanation by the providers of information led to changes in the user's knowledge, behavior, values or beliefs.

> I heard about polio immunisation on radio, but since last year's bad experience of children who died after the immunisation exercise, I had not made up my mind... I still had questions about the safety of the vaccine (unmet information needs). However, when the LCs came here, they

explained to me fully and allayed my fears that the deaths were due to malaria, but not the vaccine; I then decided to take my kids for immunisation (and showed the chart to the researcher)

Woman 6, Bushenyi

Supportive infrastructure

Provision of information alone, without the necessary infrastructure, hardly changes the situation. Women access the information but would not be able to put it into practice if they did not have the necessary resources, support from the family or the infrastructure. Health workers vividly demonstrated that issue:

The presence of safe water in the area where we conduct health education sessions about water borne disease, for example, has greatly facilitated these sessions... It is easy for the women to implement because each parish now has 2 — 3 bore holes or protected springs, and we see the effect already because diarrhoeal diseases have greatly reduced. In the past, before the Rural water project improved access to safe water in rural areas, we used to health educate, but diarrhoeal diseases were rampant!

Health worker 4, Iganga

Hence, the presence of safe water enabled women to use the information they had received from health education, which reduced water-borne diseases. Women's interest in implementing what they learned has already been highlighted as personal moderators to information access. The presence of the necessary infrastructure coupled with that interest moderates information use. It is therefore important that African Governments invest more in the basic physical infrastructure of health, including water treatment facilities, as recommended by the Uganda National Academy of Sciences (2014).

The next section presents the information activities of health workers, who are the second group of information users focused on in the book.

2.4 INFORMATION ACTIVITIES BY HEALTH WORKERS

The arrangement of this section is similar to the previous one (Section 2.3).

2.4.1 INFORMATION SOURCES

This section includes the quantified descriptive data about the actual and potential sources of information, as well as the best and easiest ways to provide information to health workers in rural areas. The analytical interpretation of sources of information then follows.

Actual sources of information

Interviewees reported as many health topics from as many sources of information as they could easily remember. The sources of health information, arranged according to the number of times mentioned by the interviewees and with the highest scoring information source listed first, are summarized in Table 2.8a.

Most important information sources

From the actual sources, health workers were asked to identify the main source(s) of information specifically for their work, and which of those were most important and why. They were also asked

---

Table 2.8a Actual Sources of Information for Health workers

Actual Sources/Channels of Information

1. Books, periodicals and treatment guidelines 2. Seminars/workshops/conferences/courses/professional meetings and other training sessions 3. Seniors and colleagues 4. Professional associations, medical bureaus and other NGOs, both local and international 5. Radio (local and international) 6. Electronic databases	7. Ministry of Health (MoH)/ District Medical Offices (DMOs) 8. Newspapers (print and electronic) 9. Television and other electronic media 10. Mobile phones/devices 11. Community-based agencies

---

Table 2.8b The Most Important Actual Information Sources for Rural Health Workers

Most Important Source	Multiple Responses	Reasons
1. Books, periodicals and other printed sources (except newspapers)	26	• For reference and in-depth information, books and periodicals are most important...they can be read or referred to when one is free and at one's own pace...Journals provide the most current literature...and printed sources are more accessible than the Internet-based resources.
2. Seminars/workshops/refresher courses/meetings and other training sessions	24	• They provide information about priority health topics in the country, and other current information in general. They also provide reports, handouts and other types of information that interviewees refer to in their work. They include practical sessions that update one's knowledge and skills. They provide an opportunity to interact and discuss health issues. They are formal learning sessions where one gets authentic information unlike colleagues—some information they provide is uncertain.
3. Seniors and colleagues	18	• Is an immediate and interactive way of getting information, it is generally free of charge and is within reach.
4. Professional associations/ NGOs/MoH and DMOs	12	• One learns new methods, skills, drugs/medicines, new diseases and other developments in the field which updates one's knowledge.
5. Online databases	9	• They provide evidence-based articles and scientific papers from the most current journal/sources, and policies and guidelines on health at national and international levels.
6. Newspapers (local and international)	3	• Although not a source of scientific literature, they provide current information, including WHO indicators on health and poverty.

to identify sources they mostly used or referred to in their work. The responses are summarized in Table 2.8b. All health workers identified more than one information source (multiple responses).

Books and periodicals

It was noted that the most important sources of information for medical doctors were books and periodicals, including the Uganda Health Information Digest (printed information) for those living in rural areas and electronic sources for the peri-urban ones, whereas those of the five TBAs and nursing assistants were fellow TBAs/nursing assistants, other health workers and seminars (oral sources). The rest of the health workers interviewed (clinical officers, nurses and midwives) identified seminars and other training sessions as the most important sources of information for their work, except five clinical officers whose most important information sources were printed materials and professional contacts/colleagues and seniors.

> We can't really do without printed sources because even during a seminar, we are given books or printed materials to read which enhance our understanding of what was discussed in a seminar.... Seminars and courses just stimulate our interest for further reading... and some seminars are quite shallow in content; so one needs to read to be able to gain knowledge and also for reference, we need books
>
> **Health worker 4, Iganga**

It was encouraging to note that health workers, mainly doctors and clinical officers, as well as two midwives, valued printed sources so much so that they bought some from their personal resources.

The printed sources that health workers mostly used in their work, including the ones referred to in critical situations, were on the following topics: drugs/treatment guides and pharmacology; childhood illnesses; medicine; obstetrics and gynecology including FP; surgery; sexually transmitted infections; nutrition; eye diseases; pathology; training of health workers and general health care. Forty-one titles were identified, of which 15 were reported having been referred to during critical situations.

The implication for information provision is that if one was to provide at least 20 titles of print materials to all health units, the 15 titles used in critical incidents would be part of the list because they have proved essential. That list, however, could be subjected to a quantitative survey to be able to generalize the findings. Furthermore, given the increasing use of mobile phones by health workers and the availability of other mobile devices, some of the basic documents could be made available electronically to ease regular updating.

Seminars, workshops and other training sessions

The health workers ranked training sessions as the second-most important actual source of information. The reasons given for that choice included being interactive, participatory, and providing an opportunity to seek clarity or ask questions, providing current information on national or local health problems, and enabling health workers to access information to update their knowledge and skills when they are out of their busy health units. Training sessions also tend to motivate health workers socially and professionally and to break professional isolation. Although six health workers reported that they rarely got an opportunity to attend seminars, they still perceived seminars as the best way/channel to deliver information to rural health workers. Comments made include: Although expensive and therefore not everybody can attend, seminars or workshops have an advantage of targeting the most common health problem and providing more focused training and current information in an environment that is conducive to learning (healthw1-Masaka)... they are participatory, so we are able to share the experiences... they take us away from our busy routines, thereby giving us a chance to concentrate on what is being taught... unlike printed information which we may fail to read because we are too busy (healthw5-Iganga)... from the seminars, I have six notebooks where I recorded seminar notes for future reference; this is the only reference material I have in my language... seminars are the only source of such information because other sources are in foreign languages (TBA).

Professionals (seniors and colleagues) and associations

Either through direct consultation and support supervision or in associations and the MoH or the District Medical Office (DMO), professionals were identified as an important and preferred source over several others. This was because personal interaction was perceived as the best way to deliver information as already reported under seminars. Other reasons that led to this choice included the fact that professional contacts provide answers as and when the need arises. There was also a challenge of a poor reading culture among some junior and/or overworked health workers, and a language limitation for the TBAs, which made some interviewees prefer professionals as an information source. For example:

> Face to face delivery of information by health workers either through visits, consultations or seminars is the best for us in rural areas where the internet is still a challenge... it should be complimented by printed information for future reference but printed information, particularly textbooks alone, should not be relied upon as a channel because they get out date. Secondly, there is generally a poor reading culture among lower level health workers such as enrolled nurses, yet some of them run health units with one or two nursing assistants, which keep them very busy... leaving very little time for them to do any meaningful reading
>
> **Health worker 3, Bushenyi**

> I find it very important to hold discussions with professional colleagues.... Reading books and/or searching the internet alone are not enough; usually after retrieving information from the web and reading, I find it rewarding to discuss what I have read with a colleague and when I apply that knowledge, we also share the experiences — what works and what doesn't, and why it doesn't work
>
> **Health worker 5, Masaka**

Furthermore, interviewees were asked whether they were members of any professional association/NGO in Uganda or abroad, and which services(s) provided by the association(s) did they find to be most relevant to their work. Out of the thirty-four interviewees, twenty-nine reported that they belonged to professional associations/NGOs, as summarized in Table 2.9. In their view, the most relevant services provided included organizing training sessions and/or conferences and other CPD activities, such as the production of documents (newsletters, reports, pamphlets, journals, posters and manuals), video tapes, radio and television programs. Most of the documents and the audio-visual materials were free of charge. The documents were described as "very useful, quite topical, informative, short articles in newsletters are easy to read, etc." The fact that the documents and activities of the associations were generally free, topical and relevant moderated health workers' information access and use and proved to be important sources of information.

Table 2.9 Summary of Health Workers' Membership to Professional Associations

Membership	Bushenyi	Masaka	Iganga	Lira	Total
Yes	8	7	7	7	29
No	2	1	1	1	05
Total	*10*	*8*	*8*	*8*	*34*

Formal and online information sources

The main focus was on access and use of formal sources such as libraries. The professional health workers interviewed reported that they depended on the small collections (20—100 titles, some with multiple copies) within the health units and on their personal documents; the majority (62%) did not have access to libraries other than the health unit collections. Only 29% (the eight medical doctors, one nursing officer from Masaka and one clinical officer from Iganga) used hospital[9] libraries, some of which were located far away from their places of work. Some health workers reported that they used more than one library. For example, Health worker 4, Masaka, used both the hospital library where he worked and Masaka district hospital library, approximately 20 km away. Similarly, Health worker 5, Bushenyi, used the two university libraries,[10] MUST and Makerere, as well as the Ishaka hospital library. Health worker 1, Lira, also reported using the hospital collection as well as the Makerere University medical library in Kampala. The TBAs (9%), however, had not used any library by the time of the interview.

It was further noted that hospital libraries provided lending and reference services to their staff, but extended only reference services to other users. When the health unit collection did not provide what one needed, health workers sought information elsewhere. The challenge was that other information units, just like the hospital libraries, only provided reference services for external users (health workers from other institutions). Some libraries, for example, at Makerere University, MUST, and Masaka hospital, had photocopying facilities, which made it easier for the external users to make copies. Lack of both photocopying and lending facilities in libraries caused frustration:

> The health centre collection which I mostly use is small... the hospital library, 50 km from here, would solve a lot of my problems but there are no borrowing facilities for us, we just sit in, read and probably take notes ... there are no photocopying services either!
>
> **Health worker 3, Iganga**

In contrast, staff working in hospitals where they benefited from all services provided by the library pointed out that:

> The hospital has an interesting collection including CME materials and the Digest. Besides reference and lending, we get information about new acquisitions which enables us to borrow new documents
>
> **Health worker 8, Iganga**

Doctors generally reported regular use of hospital library collections and the Uganda Health Information Digest that was referred to many times during the interviews; it remained one of the major sources of health information in Uganda, which had supplemented the collections in health units because it had a wider journal coverage and was more current than what was available in health unit libraries/collections. An example was given by one of the interviewees:

> the Digest focussed on Ebola before the Ebola outbreak in West Africa, and Uganda provided doctors to support the teams in West Africa... the most easily available information about Ebola was from the Digest and several doctors commented about this; and so was Hepatitis... the Digest focussed on hepatitis before the outbreak in Northern Uganda. The fact that the Digest focused on Ebola and Hepatitis, which later became topical in the region and country really made

---

[9]Hospital libraries: some hospitals within the research districts, for example, Ishaka missionary hospital (in Bushenyi), Masaka hospital and Kitovu missionary hospital (in Masaka), and Buluba missionary hospital (in Iganga/Mayuge) had what was described as a "fairly good library collection," which health workers used and had computer facilities, although the network was poor/slow.

[10]University libraries were the Mbarara University of Science and Technology (MUST) in the west, located approximately 60 km from Bushenyi, and the Makerere University Main and Medical Library in Kampala (Maklib), which is located more than 300 km from Bushenyi and Lira and more than 120 km from Masaka and Iganga.

> the Digest a key source of information. Similarly, the focus on non-communicable diseases e.g. heart and kidney diseases... health professionals have been asking for the Digest... Forensic clinical services (gender based violence, sex assault, etc) are the next focus of the Digest in 2015 as there are two aspects —health and justice- but people don't know what to do! In terms of information provision, that booklet has really made a lot of difference to Uganda's health
>
> **Doctor, Masaka**

With the exception of the TBAs, all the interviewees reported that they had attended training facilitated by Maklib staff, used the library as medical or paramedical students or external users, and/or received the Uganda Health Information Digest produced by Maklib. Some of the comments made about Maklib training are:

> The four day training I attended at Maklib in 2011 introduced me to HINARI and I am still using that knowledge and searching skills (doctor, Iganga/Mayuge); Maklib training provided me with sources of latest information on management of patients, writing proposals, etc
>
> **Doctor, Masaka**

It was, therefore, gratifying to note that Maklib had extended service to many professionals outside Makerere University, hence justifying its other role as a national reference library.

Interviewees were also asked whether they had made specific requests to libraries for reading/use in the past 1 year. Fifty-three percent of the professional health workers interviewed reported that they had made various specific requests, which further illustrated their active information seeking acts. Apart from the traditional requests to libraries, health workers also reported having requested documents from health-related national and international NGOs and associations, governing bodies such the Uganda Catholic or Protestant Medical Bureaus, as well as the MoH and DMOs.

These finding compared well with that of Brettle (2015), who reported that of the health care professionals from the northwest United Kingdom who responded to the survey she conducted, 6 had read approximately 50 evidence summaries in the previous 1 year, 7 had read approximately 25, 8 had read only 2 and 4 had read 0 or 1 evidence summary.

With regard to online sources, nine professional health workers reported that they accessed scientific health information from the Internet, four of whom used mobile modems. They all appreciated the new technologies and made several comments, for example:

> the ability to search the Internet is the biggest change in fifteen years as it has enabled health professionals to access current literature easily... we are now able to access information about new medication and to use it to manage patients better (doctor-Masaka); the major change is the reduction in reliance to printed sources as information now is more evidence based from current research and is available online
>
> **Doctor, Iganga/Mayuge**

Furthermore, six of the nine health workers who accessed online resources pointed out that in addition to using computers to search the Internet, they also used smartphones and/or other mobile devices to access the Internet and to search for any information—basic or scientific. It was noted that the Internet mobile data were getting cheaper due to competition among service providers, thus making it possible to search the Internet longer than before. Some of the comments made to illustrate the point are:

> In 1998 for example, Dr Tibbut, a retired British doctor, used to compile print copies of CME (Continuing Medical Education) pamphlets and took the trouble to post them throughout Uganda.

> The pamphlets were so useful to us young doctors then and were a major source of current and reliable information;...but now, such information is available on the Internet... I can access it using my smart phone or ipad.
>
> **Health worker 4, Masaka**

However, some other health workers pointed out that the Internet was on and off, and yet the service provider they used was expected to have the best signal among the rest of the providers. The unstable Internet network, therefore, made it difficult to download the needed articles. Hence, the Internet—access and speed—was a major factor affecting usage of online resources in rural areas, as further elaborated in Section 2.4.3 (Constraints). However, it was noted that the interviewees could use e-mails to send requests for the required articles/documents to Makerere University Library, which would greatly ease their access to the needed information. However, some were not aware of such a service and the interview raised their awareness about it.

It was also noted that the presence of collaborative arrangements provided support to access PubMed, MedScape, Google Scholar and others that were greatly appreciated. Some interviewees pointed out that when they were students, they used such resources daily for course work. When they graduated, however, the pressure for course work was over, and given the heavy clinical workloads compounded by the slow internet, the usage of online resources was no longer on a daily basis, but whenever a need arose.

As indicated in Section 2.3.4, health professionals confirmed what had been reported by the women, that the MoH had a database of contacts of all registered doctors (and other health workers) that was updated annually, and the MoH used the contacts to send text message on mobile phones about various issues on health using what was referred to as MTrack. For example, on World AIDS day, one of the interviewees shared with the author/researcher a message from MoH on his phone about prevalence of HIV/AIDS.

The media

Another actual information source was the media. It included mainly the radio and newspapers, and to a less extent television and social media. The newspapers were available both in print and online, and some health workers pointed out that they read the online newspapers, both local and international, using computers or mobile devices. The print newspapers were important to some health workers:

> The newspapers inform us about disease outbreaks in Uganda, some research break through, new drugs, jobs advertisements and advertisements for research or other funding proposals which I find very useful information that I can hardly get from any other source
>
> **Health worker 3, Masaka**

> Although newspapers are not easy to get in rural areas, they contain current information e.g. some new medical conditions especially from other parts of the world... recently I read about a congenital disease I had never heard of... a mother who delivered 8 babies and how the doctors managed the babies, some of whom were underweight, to ensure that they survive. There have also been several issues about management of breast cancer which updates my knowledge
>
> **Health worker 4, Bushenyi**

Furthermore, various local and international TV programs provided information that was considered relevant to the needs of some health workers. Others reported that they switched on particular radio stations to listen to specific programs on health that included national and international stations for current developments in the field. They were considered important sources for specific information as the comments show:

> Although Ugandan radio stations hardly offer anything new to me as a medical doctor, international radio stations such as BBC have current and relevant information which updates my knowledge on many developments in the medical field... I always tune it and sometimes I even take notes
>
> **Health worker 5, Bushenyi**

> Some radio and television programmes on health e.g. that chaired by Professor Kirya on TV are very relevant to me... some of the questions people ask the presenter and their answers update my knowledge... so, I make sure I don't miss the programme
>
> **Health worker 6, Iganga**

This agrees with the uses and gratifications theory, which posits that "the audience member makes a conscious and motivated choice among channels and content on offer... if media use were unselective, then it could not be considered in any significant degree as an instrument for problem-solving or even very meaningful for the receiver" (McQuail, 1994: 318–9).

Others reported that they listened to the radio everyday to get surveillance-related news about different topics such as epidemics, and they found radio to be an important source that supplemented the text messages sent on mobile phones about diseases outbreaks from the MoH. Preferring to use radio for what Dominick (1996) referred to as "cognition" relates to the need to strengthen one's knowledge and understanding of the current events and generally the world in which one lives.

Furthermore, eight (24%) health workers reported use of social media mainly for communication with colleagues on professional or other matters.

Community-based agencies

These included faith-based organizations and the LCs. They were the least mentioned actual sources in Table 2.8a, mainly because they did not provide scientific information as traditional information units did, but they facilitated access to such information in various ways. For example, approximately 40% of the health workers interviewed reported faith-based organizations, mainly the church, to have supported information provision either directly by buying information materials or indirectly by organizing and/or sponsoring seminars for TBAs and professional health workers:

> I work for the church diocese and it sponsors me sometimes to attend seminars (healthw2-Iganga)... the church pays for the newspapers and some reading materials for the library which provides me with information
>
> **Health worker 3, Bushenyi**

Furthermore, the church, like the LCs, was considered important in communicating different types of information to the public, and therefore assisted health workers in public health information dissemination and health promotion.

Although the role of LCs in providing scientific information to health workers was negligible, the LCs were reportedly very important in identifying disease outbreaks and other health problems and in reporting them to health workers, health units and/or the DMOs. The LCs also facilitated health workers in health promotion by mobilizing communities for health activities and by enforcing some health issues such as hygiene and sanitation, as indicated in Section 2.3.4 about women.

Potential sources of information

The actual sources of information presented had remained potential sources to some health workers who were not able to access and use them. However, such health workers identified the sources as important, even though they had not used them or gotten information directly from them yet.

Best and easiest ways to access health information

The subsection presents what health workers perceived and experienced as the best and the easiest way(s) to access information (summarized in Table 2.10). Like the women, it was noted that the best way may not be the easiest, and vice versa. Best, as expressed by the interviewees, had connotations of value, whereas easiest had links with overcoming constraints. This also highlights the relationships between needs for information, value of information/source, constraints and moderators. A part of this subsection was about women: health workers reported what they considered to be the best and easiest ways to provide information to women, as indicated in Sections 2.3.1 and 2.5.1.

It was noted that printed information was identified both as a best and an easiest channel; the main distinction was in size. Whereas smaller pieces of information such as circulars, handouts and pamphlets were easy to photocopy or scan and distribute to as many health workers as possible, and although health workers found them easier to read, larger pieces such as reports and books, although expensive, were identified as the best way to send detailed or in-depth information. Printed sources in general had several advantages: when sent out, more health workers would get them than those attending seminars. They could be used by many and could be referred to whenever the need arose. However, the interviewees who had stable Internet access preferred documents sent to them as an e-mail attachment, whether small or large. Several others identified the easiest way as sending text messages on mobile phones, which the MoH was already doing.

Table 2.10 Best and Easiest Ways to Provide Information to Rural Health Workers	
Best Ways/Channels	**Number of Health Workers Interviewed**
Reports, books and other (large pieces of) printed information	17
Through training, eg, in seminars/workshops/refresher courses	11
Electronic as e-mail messages or attachments	6
Easiest Ways/Channels	**Number of Health Workers**
Pamphlets/handouts/circulars (small pieces of printed information)	14
As text messages on mobile phones	12
Through fellow health workers (eg, support supervision, liaisons, meetings, job training)	5
Radio	3

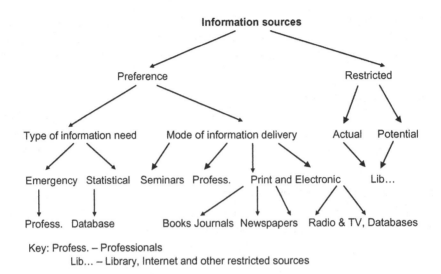

FIGURE 2.3

Analytical interpretation of health workers' information sources.

Other analytical subdivisions of health workers' information sources

Analysis and further interpretation of information sources revealed four subcategories: preference; restricted; information behavior which are presented in this section; and constraints to information access presented in Section 2.4.3 (Constraints). In addition to what is illustrated in Fig. 2.3, preference and restricted, have a lower-level property value of information source, which is discussed in Chapter Three (section titled Value of Information—Clinical Work). Furthermore, the choices of or preference for information sources elaborated in this section also show that interviewees valued those sources. However, interviewees identified several sources that, due to a number of constraints, remained only potential sources of information, and these formed another branch of restricted sources. The fourth and last subcategory presented in this section is information behavior.

Preference

Unlike restricted sources, health workers made a choice regarding which sources they preferred. The choice was generally dictated by the type of information needed or by the mode or method of delivery of an information source as elaborated in this section.

The choice of information source was influenced by the type of information needed. When health workers needed information in emergencies, for example, they preferred certain sources to others. Similarly, when they needed local statistical information, the district database or the MoH—Health Management Information System (HMIS) were considered the best and preferred choice. The type of information need, therefore, was subdivided into emergency and statistical.

In many emergency or critical situations, the efficiency of an information source in terms of specificity and speed of delivery of a response was the overriding factor. For example, in labor emergencies, midwives reported that they prefer professional oral advice to print information because of its specificity, speed, and the fact that one can interrogate the source and get an immediate response as the comments show:

> A maternity emergency is so critical that it does not allow one to look for information in books... there is no such a time as reading books... in such a hurry and with a woman crying, I doubt whether I would even understand what I read. So, I consult my seniors by phone or during day I call for my boss, the health centre in charge, and when I do so, I need a definite advice from her, if I don't get it, I refer the woman to the nearest hospital... It is after referral that I get my books and obstetric manuals and read about that particular condition so that when I get it again, I would know what to do... This is how I have gained the experience I have
>
> **Health worker 1, Bushenyi**

For statistical information, the choice of the source was made because of its efficiency in delivering information quickly (information was printed from the database rather than retrieving it from paper files in some other offices) and its effectiveness in terms of the quality of information.

> When I need information about population and related issues, I go straight to the District Population office database because this is the best place in the district where one can get such information from — the figures are reliable and I get what I need quickly without having to go through misfiled paper documents... Other offices also get such information from there; so, it is better to go to the primary source and get uncontaminated information
>
> **Health worker 3, Bushenyi**

With regard to the mode of information delivery, the choice of information source was affected by the manner in which information was delivered as well as the perceived usability of the source. As Table 2.8b shows, most health workers preferred books, periodicals and other printed sources, but seminars/workshops/training session, fellow health workers and the media, in that order, were also important.

Restricted

Although the examples given indicate that health workers could choose which information sources to use, in some other situations there was limited choice, thereby making such sources actual but restricted. Some information was not available and had to be searched and/or requested from medical libraries that had a wider and deeper coverage of medical/health topics. In this subcategory, there are sources that were used, although rarely (actual), and those that remained potential sources of information. Library and online information resources, for example, were actual sources for some interviewees but potential sources for others, as already indicated.

Specific information needs that could not be satisfied by the available sources (eg, professional colleagues, books and periodicals in personal and/or health unit collections) made health workers seek information elsewhere. The needs required sources—print and/or electronic—that could only be accessed in medical libraries or online databases:

> Since I joined the profession, Uganda had not had cholera... when we recently got an outbreak of cholera in the district, I had last read about it when I was a medical student 4 years ago... so, what I did was to read my textbooks which were not satisfactory. I enquired from colleagues who, unfortunately, were as ignorant as I was. I then went to the nearest library, 40 miles away at MUST; luckily, I found some textbooks on tropical medicine and I searched PubMed which

> provided relevant and interesting information that updated my knowledge and answered the questions I had satisfactorily
>
> **Health worker 35, Bushenyi**

> The hospital where I work doesn't have a library, but a small collection which we, doctors, can borrow from, although it is quite limited. So, whenever I go to Kampala, I have to create time to use the School of Public Health and Albert Cook libraries for whatever needs I have, but I can't borrow books. Fortunately, I can do literature searches and print the abstracts, or if I want an article from a journal, or something from a book, I can make photocopies
>
> **Health worker 1, Lira**

Furthermore, four health workers who had used the Makerere University Library document delivery service (DDS) before the interview reported the difference it made in their information needs because they only sent e-mail requests rather than making journeys.

However, health workers who were not able to travel to medical libraries or to access online information resources and were not aware of the document delivery facility reported such sources as potential. This is discussed further in Section 2.4.3 (Constraints to Information Access).

Information behavior

Health workers interacted with sources actively or passively to access information. The actual sources of information used are outlined in Tables 2.8a and b. The information behavior included reading (print and electronic sources), consulting (fellow health workers), attending training sessions (seminars, workshops), searching (electronic databases, library collections), listening (to radio), and, to a limited extent, watching (television). Section 2.4.2 discusses information behavior in more details.

2.4.2 INFORMATION NEEDS

This section highlights information that had not been accessed, what made health workers need information most frequently, the most difficult type of information to access, information that had been accessed and its description, the analytical interpretation of data about information needs, and behavior.

It was important to find out, at first, whether the need for information existed and, if it did, if it had been met. All the health workers interviewed reported that they have had various needs for health information. However, the needs of most health workers had not been fully satisfied. That was mainly due to the continuous changes in the medical field, which make health workers need information to keep abreast of the new developments. It was, however, compounded by the limited access to the required information/sources. The summary of responses is as follows:

- Health information needs:
- Not satisfied: *15* interviewees (44%)
- Partially satisfied: *10* interviewees (30%)
- Fully satisfied: *9* interviewees (26%)

The information needs of 44% of the health workers interviewed had not been fully satisfied, whereas 30% reported that their needs were partially satisfied. The rest (26%) had their needs fully satisfied, and these were the health workers who had a stable Internet access that enabled them to address their information needs fully.

Table 2.11 Topics for Which More Information was Needed by Health Workers	
Topics	
1. Reproductive health (including FP)	11. Surgery
2. HIV/AIDS	12. Psychiatry
3. Health: various aspects	13. Adolescent health
4. Malaria	14. Heart diseases
5. Education and training	15. Diabetes
6. Childhood illnesses	16. Eye diseases
7. Drugs/medicines and vaccines	17. Ebola
8. STIs other than AIDS	18. Leprosy
9. Tuberculosis (TB)	19. Trypanosomiasis
10. Government policies on health	20. Theater procedures

Information that was not fully accessed

Health workers' unmet information needs and the topics they needed more information about are summarized in Table 2.11, with the topic that scored highest listed first and the rest in a descending order.

All the health workers interviewed needed more information for updating knowledge and for the management of patients in general, but some indicated specific conditions, diseases or health issues, for example, management of rare cardiovascular diseases, current management of diabetes, pediatric psychiatric disorders, management of skin disorders, Ebola and the new HIV/AIDS drugs. Other needs, which were not common to all, included information about training and/or research opportunities and production of documents, community support, administration, preventive care, as well as health and gender relations.

Furthermore, the most difficult type of health information to find or access was identified to be current literature, particularly by health workers who were not able to access online resources. Lack of various types of current information, therefore, left health workers' needs unsatisfied, as indicated at the beginning of this section.

Information that had been accessed

Health workers generally reported that they had accessed some information on various topics or diseases and from various sources. Of the thirty-four health workers interviewed, 66.7% perceived the information so far accessed as satisfactory and relevant and that it updated their knowledge. However, 9.6% pointed out that some of the information accessed did not satisfy their information needs fully, mainly because it was out of date, the sources or channels of information were neither appropriate nor satisfactory, and information about some topics lacked the required details. The unsatisfactory quality of information accessed left health workers' needs unsatisfied on a number of topics, thereby (the quality of information) becoming a constraint to information access or use. The comments are summarized in Table 2.12.

Health workers made various detailed comments about the information they had accessed. Some of the comments are presented in Section 2.4.3 (Constraints), whereas others are in Section 2.4.1 (Information Sources); some in this section and the satisfactory ones are highlighted in Chapter Three as value of information. That shows the linkages between the categories that comprise the model presented in Chapter Four.

Table 2.12 The Description of Information Accessed by Health Workers	
Health Workers' Perceptions of Information Accessed	**Percent**
Satisfactory/relevant/updated knowledge	66.7
Fairly good but some information was missing	23.7
Not satisfactory/lacked the necessary details	9.6

Other analytical subdivisions of health workers' information needs

In trying to interpret the data further, it was noted that while the information needs category for women (Section 2.3.2) was subdivided into latent, critical and active, it was different for health workers. Most of the health workers' needs were apparent and led to active information seeking In some of the examples given to illustrate the concepts, interviewees' needs for information were difficult to separate from information seeking acts. Furthermore, the need for or the reason why an interviewee decided to seek information, the strategies used and the sources used to get the information, the purpose the information will serve, and/or the use to which it is put when received may all be in one quote, especially those recorded from critical incidents. That was particularly true for the met information needs; the unmet needs were relatively brief.

Critical incidents could actually be considered within the value of information category because they reflect on how information solved emergency or critical needs. However, because the focus of this section is to highlight the various needs for information, critical incidents are presented within this section, and only a few examples are cited in Chapter Three.

Within the apparent needs and active information seeking, the main issues that emerged from data analysis were: clinical information needs, education and training, preventive care, and causes. It was further noted that the information needs were inter-related, for example, an interviewee would need further education to improve patient care (clinical) and provide more preventive care. Critical information needs and the need for reference information were found in all the subcategories. For example:

> I need information to refer to in training nurses, in writing research proposals and some reference materials on research methods *[education]*; I also need information on current management of paediatric illnesses and in emergencies *[clinical]*
>
> **Health worker 1, Masaka**

> I would like to read about the new Government policies on health, e.g. gender and health, to refer to in my community work of sensitising and mobilising people about the control of various health problems *[prevention]* and some reference information to explain the high rates of anaemia in children *[causes]*
>
> **Health worker 3, Bushenyi**

Furthermore, the need to keep up to date was common to all interviewees, as already indicated, and was found in all the subcategories, although it is primarily discussed under continuing education, usually referred to as continuing professional development (CPD). Similarly, information needs for research could not stand on their own as a subcategory or a property of one subcategory because they were found in all the subcategories. Fig. 2.4 presents diagrammatically the health workers' information needs subcategories and properties.

FIGURE 2.4

Analytical interpretation of health workers' information needs.

Clinical information needs

Information needs related to clinical work were reported by all the interviewees except two who were not practicing clinicians. Clinical information needs included mainly treatment and delivery of health care. Within that subdivision, several issues emerged that were considered difficult or unusual, and there were also the usual or ordinary information needs. The unusual needs resulted from emergency situations reported in critical incidents. Although they may be regarded as unusual, they were generally the basic information needs in the professional activities of health workers.

Treatment and delivery of health care information needs referred to the (diagnosis and) treatment of patients and management of emergencies or referrals, including the maternal and related emergencies. Some of the health workers' needs for information were classified as usual, whereas others were unusual. The usual information needs for treatment focused mainly on malaria, HIV/AIDS and other STIs, childhood illnesses, ulcers, diabetes, skin diseases and drugs/medicines in general. Some of the clinical information needs led to constraints to information use reported as unavailability of medicines/facilities in Section 2.4.3. Many health workers (58%) had needs for information concerning management of resistant malaria. Several others (23%) needed information about new drugs and management of AIDS patients and effective medication for various illnesses. An example:

> I have had several cases when I want to prescribe medicines for example, for ulcers and find that the patients had already used the medicines I know, but didn't improve; yet I don't know what new or alternative drugs are available. . . so I search international sources, which give a range of medication, but on checking, I find that we don't have such medicine, so I have to search again which is frustrating!
>
> **Health worker 1, Masaka**

Unusual treatment information needs were reported on various topics such as malaria, childhood illnesses, trypanosomiasis, onchocerciasis, surgical emergencies and a rare type of fever that was still undergoing investigation by the MoH. The needs become even more critical when patients decline to be referred, or when there are genuine transport problems involved in taking the patients to a referral hospital. That leaves the health workers with no option other than to struggle to solve the emergency. Some examples:

> Last month, I gave antimalarial to a female patient who reacted to it after a day. When she came back to see me, I instructed her to stop taking the drug and I gave her some treatment to

counteract the reaction... For some strange reasons, however, the woman continued taking the drug. She was brought back after a week when she was badly off and could not even walk. I asked her whether she continued with the drug, and she said 'yes'; why did you continue - no answer! I decided to refer her, but she and her relatives refused... I then admitted her... In my working experience of 9 years, I had never gotten such a case... normally when I tell patients to stop the drug, they do so... Anyway, I read BNF about adverse drug reaction and followed what was recommended. Within 2 days, she had improved and I discharged her on the third day. In the process of taking the drug, malaria got cured... I tested her, and she was negative

Health worker 4, Iganga

A female patient was brought unconscious to the health centre. After some treatment, she gained consciousness but later lost it again. I wanted to refer her but it was late at night and transport was not readily available! I was puzzled... but from the history, I strongly suspected it was trypanosomiasis. I checked the National Standard Treatment guidelines because I thought the treatment I had given was not appropriate. I found out that I had to give another type of drug in addition to what I had given ...when I did that, the patient improved greatly and was later discharged

Health worker 5, Iganga

It was noted that in critical situations, health workers mainly used the available information sources, treatment guidelines, or other reference sources to assist them in the diagnosis to confirm or not confirm what they had suspected and to check for the recommended medicine and/or the appropriate doses of the medicine they planned to prescribe. Others referred to available documents for both diagnosis and prescriptions. In some cases, health workers reported having referred to several documents:

A 2-3 year old child was brought unconscious after intoxication, but there was no proper history... I referred to my emergency handbook 'National Standard Treatment Guidelines' but it was not definite, I cross checked the 'Medical practice in developing countries', this was no better; I then checked the 'Intoxication management' manual and got the information I needed... This enabled me to manage the problem

Health worker 7, Iganga

In other instances, mainly surgical, doctors reported having referred to evidence-based literature and advice from colleagues and seniors to assist them in performing such operations, sometimes for the first time. In other cases, the available information sources and colleagues failed to provide the required information and the health workers resorted to referrals. Generally, the doctors and midwives who handle gynecologic and obstetric cases had experienced more unusual than usual information needs.

There were also other general clinical information needs that combined treatment/management and delivery of health care as well as counseling of patients.

Education and training information needs

These included CPD, formerly referred to as continuing medical education (CME), training of health workers, production of documents and postgraduate information needs.

With regard to CPD, three types of information needs emerged from the data: updating, retraining and other. The need to advance one's knowledge and to keep up to date with new trends or developments in the fast-growing field of medicine was so central that it emerged in all the subcategories and was inevitably highlighted in other sections such as information sources, constraints and moderators. Furthermore, the need for updating was reported by all the health workers interviewed.

Besides updating per se, there were needs for retraining or further education, but it was difficult to draw a line between information and training needs in some situations. The need to learn what one had not learned clearly emerged from the data. This was because either an interviewee was doing work she/he was not trained to do, such as a midwife working as a nurse, or the initial training (mainly in midwifery and nursing) did not prepare such health workers for the work they were doing at the time of the interview. The needs were also due to advances and changes in medical knowledge and practice. Such training and information needs were mainly identified by enrolled nurses and midwives:

> I was only trained as an enrolled nurse, but I am now doing midwifery work at the health centre, of course with the assistance of the health unit head who is very experienced. So, I need to attend a course to enable me to gain knowledge about midwifery issues
>
> **Health worker 1, Bushenyi**

> I need a refresher course to ease my work, for example, I was trained over 20 years ago, and we were not supposed to use stethoscopes unlike students of today... the situation now dictates that we attend to various cases
>
> **Health worker 8, Masaka**

Clinical officers and doctors also highlighted certain topics that were related to inadequacies in their initial training, such as:

> Our training was not detailed as far as nutrition is concerned; so, I need further training on nutrition and related topics and I keep reading about them
>
> **Health worker 4, Iganga, clinical officer**

> the medical school training in psychiatry was not deep enough for us to be able to handle the various cases we come across in practice e.g. paediatric and HIV — related
>
> **Health worker 1, Masaka, doctor**

Some "other" education or training information needs that were not strictly updating or retraining included capacity building and/or skills development to implement programs or projects. During the interview, however, it was not possible to draw a clear line between an information need and a skills/training need. Specific skills were needed in management, communication, counseling and life skills.

> I am in charge of a community based programme but I have never trained in management, budgeting, etc. I read some books but I also need a short course... I have seen some relevant courses advertised in institutions in Kampala as evening courses but the problem is that I am too far to attend; a full time short course would suit me better
>
> **Health worker 7, Iganga**

> I need communication skills to perfect my public health information delivery in the rural outreach health education programme. I understand such workshops are run in the district, so, I applied to the DMO recently and wait for a response
>
> **Health worker 4, Iganga**

> I need skills to be able to assist and educate adolescents and parents about Life skills e.g. self esteem and negotiation skills... to equip young adults with skills to enable them to go through adolescent problems successfully
>
> **Health worker 9, Bushenyi**

The findings have implications for the education and training of health workers in Uganda, which will be detailed in Chapter Five.

There were also needs for information to assist health workers in the production of documents such as reports, conference/seminar papers, and research proposals, and for publishing. Furthermore, some health workers reported writing conference or seminar papers and research proposals as part of their critical incidents because they got to know about the conference/seminar or proposals late; so, they had to search for the necessary information, which was not easy. The comments here illustrate such information needs:

> Recently, I was invited to present a seminar paper about maternal and child health, but I got the invitation late. I needed information on the topic in general, but specifically on maternal and infant morbidity and mortality which was difficult to get by year, age and at parish or sub county levels. I checked at the district population office and at the DMO but could only get part of the information I needed. I consulted a medical doctor who recently conducted a seminar in the district (Bushenyi), she provided some information which assisted me to write the paper. However, the paper did not include information by parish or sub county because it was not available
>
> **Health worker 6, Bushenyi**

Information concerning postgraduate studies was also needed by some health workers, particularly distant or correspondence courses that would enable them to study as they continued working.

Preventive care information needs

These included information to support health education or promotion in the community and other preventive information needs. For example:

> I need facts and figures about adolescent health problems (HIV/AIDS and other STIs, pregnancy, drug abuse, etc.) nation-wide to enable me to run the school health programme (in schools and communities) more efficiently
>
> **Health worker 3, Bushenyi**

Preventive care information needs other than health education and promotion included prevention or control of diseases and other health problems. For example, several health workers needed information about HIV/AIDS vaccines, malaria control and surveillance.

Causes

Finally, it was noted that health workers' needs for information about causes of diseases or health problems were not an end in themselves, but they were a means to prevention. The comments here illustrate that point further:

> I would like to know what has caused the sudden increase in TB cases... is it due to reduced immunity or resistance to drugs?... I have not been able to find this particular information although I know the usual causes of TB... we have to control the situation
>
> **Health worker 3, Masaka**

> I have been trying, without much success, to find out the cause of high incidences of paediatric anaemia and convulsions in the area served by this hospital... I consulted colleagues who have worked here longer, but they don't know why it is high although, they too agreed that there are many such cases... I may have to do some mini research to find out why
>
> **Health worker 8, Iganga**

> I need to know the circumstances that lead to mental problems in children especially if they have anything to do with the type of deliveries we (TBAs) conduct in the villages, and how they can be avoided
>
> **Health worker 10, Bushenyi**

Information behavior

Analysis of data concerning the need to know inevitably led to information behavior and whether the information needs were satisfied or not. As already pointed out, health workers easily accessed some information passively, but the main method of information acquisition was through active seeking by reading printed or electronic sources, searching the Internet or local databases, attending seminars and other training sessions, holding discussions or consultations with seniors and colleagues, and with professional associations. In some cases, listening to radio and watching television and other media were also part of active seeking. The information seeking acts are also highlighted in the sections on information sources as well as personal and interpersonal moderators, because it would be difficult to discuss information behavior without relating it to sources, needs for information and its role in moderating constraints.

Active information seeking

In response to their information needs, health workers used several strategies to access information. In urgent or critical situations, for example, health workers actively sought information from seniors and colleagues, books/journals/manuals, online sources, seminar/course notes and/or library/health unit collections.

With the exceptions of the TBAs who, in critical situations, either referred the patients (to higher level health units) or called in a fellow TBA, of the doctors who worked in hospitals where there were other doctors or professionals to consult or who could search online sources, and of the midwives who pointed out that they frequently consulted their bosses because of the nature of obstetric emergencies, all the other health workers interviewed pointed out that they started by reading the available printed sources. When such sources do not provide the needed information, then the health worker would consult colleagues. The colleagues they consulted were usually from the nearby health units, rather than from within (the health units), because many of them were more senior than the people they worked with.

The health workers, who included the library in their information seeking activities, clarified that reading was also the first step. Six of them pointed out that they used smartphones or other mobile devices to search the Internet and read. Others read printed sources from personal collections, also as the first step, followed by consulting colleagues or seniors; using library resources was the third step. Related to library use were the electronic sources that were from the HMIS and/or the District Population database, whereas four reported sending literature search requests by e-mail to Makerere University Library.

In situations other than the critical or urgent ones, reading was also a major strategy. Apart from the TBAs, the rest of the interviewees reported that they read printed and some electronic sources to be able to keep up to date with the developments in the medical/health field, although it was generally difficult to access current literature, as already pointed out. Unlike some enrolled nurses and nurse aides who reported that they rarely read, some other health workers, mainly doctors and heads of health units, reported reading something daily. This was interesting because they are very busy people, but the value of information seems to have compelled them to read:

> I must read something everyday, be it a book, an article, report or magazine on health... reading is my first choice when I need information... I also consult colleagues but I do read everyday... in fact, sometimes I contact colleagues for reading materials
>
> **Health worker 7, Iganga**

All health workers interviewed pointed out that whenever the need arose, they sought advice, held discussions, or consulted their seniors or colleagues. The consultations were mainly face-to-face and/or by telephone because all the interviewees had mobile phones. Similarly, all health workers had attended seminars and other CPD sessions. Health workers generally reported that, most of the time, they used a combination of strategies to be able to get the information they needed:

> I read books and other printed sources in our health centre collection but I need more current stuff...Recently, I attended a one month workshop on child health, and a 6 week course on FP, and I also learned how to request for literature from Makerere University library. Reading printed and electronic sources and attending seminars are complementary sources of information for my work
>
> **Health worker 3, Masaka**

> I contact the DMO office for various records and reports, refer to hospital records and reports, read books and periodicals, CME materials, attend seminars and doctors' meetings, I download articles from the internet, with difficult though, and I hold discussions with colleagues whenever need arises
>
> **Health worker 7, Iganga**

The findings confirmed that reading and consulting or talking are important information seeking activities for rural health workers. The findings also revealed that for both women and health workers, the more educated an individual was, the more actively she/he sought information. Medical doctors actively sought information, whereas most enrolled nurses passively waited for handouts from their seniors, and some confessed that they read when they got a complicated case but regularly consulted colleagues and seniors.

Passive access

To a lesser extent, health workers reported that they acquired some information passively by chatting with colleagues and seniors informally, listening to the radio and/or leisurely watching the television. The main information behavior as indicated was active seeking.

Although the book had not primarily focused on information seeking behavior as such, and although its concern was in regard to information behavior in general, it has been able to highlight the health workers' information seeking activities that can be used to improve health information provision in Uganda.

2.4.3 CONSTRAINTS

This section presents the descriptive data as well as the analytical interpretation of data about factors affecting health workers' access to information. The factors included both the negating and supporting ones, whose descriptive data are presented in this section to ease comparison. The negating factors were interpreted as the constraints and are presented in this section, whereas the supporting factors were the moderators. The analytical interpretation is presented in Section 2.4.4.

Table 2.13 Summary of Factors Affecting Rural Health Workers' Access to Information	
(a) Enhancing Factors	**(b) Negating Factors**
1. Personal attributes	1. Limited access to information/library/Internet
2. Holding training sessions in rural areas	2. Economic hindrances
3. Availability of online databases	3. Poor IT infrastructure
4. Support supervision	4. Shortage of staff
5. Use of mobile phones	5. Late invitation to training sessions
6. Motivation	6. Lack of/limited computer skills
7. Nature of the profession	7. Geographical location
8. Interpersonal factors	8. Apathy/passivity
9. Facilitation and communication	
10. Media rural outreach	

Enhancing or supportive factors

Similarly, rural health workers identified various factors that had enhanced their access to health information. The factors are summarized in Table 2.13a. They also proposed ways in which such factors could be strengthened, for example: the need for Government to devote more resources to CPD and improved provision of information to rural health workers.

Negating or hindering factors

Health workers identified factors that had hindered their access to health information, as indicated in Table 2.13b. They made some suggestions about how the factors could be overcome, such as the need for DMOs to publicize the information facilities available in the district as well as the planned seminars/training opportunities for rural health workers to know in advance and be able to plan accordingly, and the need to monitor whether the information materials distributed reach their destinations. Health workers also identified factors that had hindered rural women's access to health information, as reported in Section 2.3.

The analytical interpretation of negating factors indicated that there were two types of constraints that emerged from the data: constraints to information access and constraints to information use, as Fig. 2.5 shows.

Constraints to information access

All interviewees acknowledged the general improvement in information provision but pointed out that the remaining challenges needed to be addressed. As indicated in Section 2.4.1 (Information Sources), some health workers had to struggle to be able to access the needed information, but sometimes with minimal success. The findings further revealed that, in some situations, several constraints combined to hinder health workers' access to information.

Constraints concerning specific information sources

An overview of information sources used by the interviewees is presented in Section 2.4.1. The difficulties experienced in accessing some sources are highlighted in this section.

Limited access to medical libraries and to relevant and current printed sources featured high among the factors that had hindered some health workers' access to information, as already presented in Section 2.4.1.

With regard to Internet-based information resources, some health workers pointed out that there was too much information on the Internet, which made it time-consuming as one spent time trying

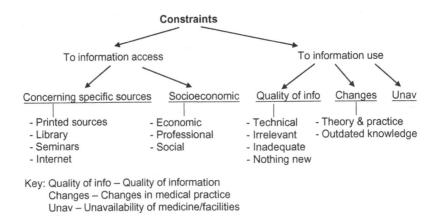

FIGURE 2.5

Constraints to information access and use by health workers.

to get what was needed and yet the Internet speed was slow. It was also noted that in many rural areas, the Internet signal was poor, which made it difficult to benefit from the gradually decreasing cost of Internet mobile data. Limited searching skills were another constraint, mainly among the older and lower-level health workers. Some comments:

> As an individual, who went to the medical school sometime back, I was used to print but now, I know there is some good information on the web, but when I try to get it, it takes a lot of my time... this change has come with a cost on our time and some of the information is not applicable here, it is too high tech
>
> **Health worker 7, Iganga**

> Due to heavy work load, some health professionals I work with fail to get time to search the internet for literature or to practice literature searching and they get discouraged further by the slow internet...The cost of airtime on the modems for searching literature is limiting especially for those who are not experienced in searching as they spend time and yet the internet is slow... Several ordinary lower level health workers I have interacted with did not know how to search the internet, and they complain about internet being too expensive
>
> **Health worker 8, Iganga**

> The internet in many areas where I do support supervision is still poor... even though the cost of the internet mobile data by various service providers continues to go down, health professionals in the affected settings are not able to benefit from the reduced costs even with smart phones
>
> **Health worker 1, Masaka**

> There are many positive changes, for example, now I have an email account, which I didn't have twenty years ago and I use it to send and receive information directly, but at a cost... I go to the internet café and the internet speed is bad... the hospital has a modem but it is slow, the café is slightly better... so, I value the updates from training courses and workshops and printed sources such as the Digest
>
> **Health worker 1, Lira**

Concerns about seminars and other training sessions had several aspects, such as lack of funds and/or failure to attend the session because of staffing shortages, which are discussed in the subsections about economic and professional constraints, respectively. Other issues included limited publicity, late invitations and selection procedures that resulted in some would-be attendees missing the training sessions.

Socioeconomic constraints to information access

They are subdivided into economic, professional and social, although there are some inevitable interdependencies between them. As Table 2.13b shows, some of the most important constraints are economic, followed by professional and then social. The economic constraints included financial limitations and technological and spatial factors. Generally, the economic constraints tend to underpin other factors, hence the interdependency. Technological constraints, for example, were very much related to financial factors as already indicated.

Financial limitations affected sources of information, as well as the morale of health workers. The most affected sources were the Internet-based resources as well as the current books and periodicals, as already indicated. This was mainly due to the economic situation in Uganda as a low-income country, and one of the interviewees commented:

> The problem is how to divide a small financial 'cake' among the many competing needs
>
> **Health worker 3, Iganga**

Furthermore, it was noted that financial limitations generally constrained attendance of conferences and courses more than seminars. This was because it is cheaper to sponsor individuals to attend seminars than courses and international conferences. Some health workers tried, without success, to get sponsors, as indicated in the comment below:

> I get to know about workshops and conferences which require registration fees, etc, but I fail to attend because I can't afford. Sometimes, I look for sponsors but don't succeed. I have also seen several relevant courses advertised in international newsletters, but I don't have funds to attend
>
> **Health worker 3, Masaka**

The effect of financial constraints on the morale of health workers was also highlighted. Senior health workers and heads of units pointed out that low morale among health workers was a serious problem that affected the various aspects of their work, including information acquisition. It was also noted that distance affected access to DMOs, libraries and training institutions, as well as access to other information sources, particularly where the IT infrastructure was poor.

The main profession-related constraints identified were shortage of staff and/or time, and to a lesser extent apathy and passivity. Although profession-related, some of the constraints were aggravated by the economic factors, for example, staffing shortage in health units. Understaffing leads to a heavy workload, which makes health workers too busy to spare time to seek information as they would have wished, for example, to read more, attend seminars/courses and perform research:

> On several occasions, I have failed to attend seminars where I am invited because one of my staff may be on leave and the other may be away, e.g. on a course, or sick... I can't close the unit; so I miss the seminars because of the staffing problem
>
> **Health worker 4, Iganga**

> Under-staffing creates a lot of work for us... it is difficult to find time... I have been planning to do some research, but I have failed... there is also a correspondence course I

> have been planning to take, but that means adding more work to what I have; so, I keep postponing
>
> **Health worker 3, Lira**

Apathy and passivity were also identified as constraints to information access. Generally, these were observations made by the heads of health units, who also observed that some lower-level health workers had poor reading culture and poor IT skills, whereas others pointed out that some of their juniors were passive:

> Lack of exposure tends to make health workers passive e.g. if one doesn't attend seminars/courses to appreciate the information accessed and to be challenged or motivated by the experiences of other seminar participants, one may become apathetic about information seeking... Hence apathy or passivity are only symptoms caused by lack of exposure
>
> **Health worker 6, Bushenyi**

From this, it appears that apathy or passivity may be related to a combination of factors, namely, poor reading culture, poor IT skills, lack of exposure and higher aspirations. The factors seem to have constrained some lower-level health workers' information access and made them less active information seekers.

Social constraints, however, were mainly about security issues. Insecurity and the resultant lack of peace were reported mainly in Lira district due to the war in Northern Uganda. The factors interrupted the day-to-day activities of rural people, including health workers, thereby hindering their access to information.

Constraints to information use

As indicated in Section 2.3.3, the fact that there is a need for information and the sources of information are accessible is no guarantee that the information contained therein will be used. It was noted that although some information was accessed, a number of constraints intervened to stop its use. This was mainly due to the quality of information accessed. In other situations, it was reported that information use had led to changes in the user's knowledge, but that knowledge could not be used/applied. This was mainly due to the contradictions between theory and practice, and the unavailability of some medicines and other facilities such as laboratories in lower-level rural health units. Finally, health workers pointed out that they could no longer use some of their knowledge because it was outdated. Hence, constraints to information use were subdivided into three groups: quality of information, which was the major constraint; changes in medical practice and unavailability of medicine and facilities. It was further noted that the economic situation in the country gave rise to some of these constraints.

Quality of information accessed

This subsection and the previous one about constraints to information access confirm, among other things, that information access and use in rural Uganda faced challenges of both quantity and quality of information. Some lower-level health workers (enrolled nurses and midwives, nursing assistants) pointed out that some of the information they accessed was not understood because it was too technical and/or too advanced, and some was inapplicable in rural health units:

> Some of the books and periodicals we have here (at the sub county health centre) are too technical... some include very advanced equipment (this was in relation to management of heart

diseases) which are not available here... so, the information in the former is not understood, while that in the latter is not applicable

Health worker 8, Bushenyi

I contact various organisations and request them to send us documents... and they do. However, some of the books they send contain information which is too technical...really meant for doctors not us... Sometimes, I take a book with an interesting topic to seminars and ask doctors to clarify or explain some things which I don't understand and that help

Health worker 3, Masaka

Irrelevancy of the information received was another constraint. This was very close to the previous one in regard to applicability, for example:

The information in some of the textbooks we have about paediatrics, public health, internal medicine and pathology is not very relevant to our current tropical health situation because they were written in the West... the focus is not tropical medicine

Health worker 5, Bushenyi.

I receive the Lancet and BMJ, though irregularly, and I find them very good for updating my knowledge, but generally about 80% of the articles are not relevant or applicable to Uganda's situation especially to a doctor practising in a rural hospital... if one is a lecturer in a medical school, it should be relevant

Health worker 4, Masaka

The fact that information is not applicable to local situations renders it irrelevant to that particular user and in that particular situation. The same information could, however, become relevant if that user takes an academic/research post in a medical school or research institution. Hence, the journals referred to in these quote are relevant to some Ugandan health workers but not to others.

Inadequacy was also identified as a constraint. The type of information referred to was relevant but inadequate in content because it lacked details and failed to provide the information health workers needed:

Some project reports are quite inadequate in content... they lack the details a medical personnel would need... This could be due to the fact that the authors are trying to cater for a wider audience, but it doesn't stop us from feeling that we are not catered for!

Health worker 1, Lira

Closely related to inadequacy was the information that was reported to have provided "nothing new" to health workers. This referred mainly to the basic public health messages on radio and television, as well as the repetitive topics in seminars/workshops.

Changes in medical practice

Failure to keep up to date with the continuous changes in medical practice led to differences between theory and practice and to "outdated knowledge," which constrained information use. Unlike the "quality of information" constraint, in this subsection health workers pointed out that although the information they had received led to changes in their state of knowledge, they could not put the knowledge into practice because of the changes (theory vs practice constraint). This also raised the issue of reality versus ideal as the comments show:

> The doses of some medicines as recommended in the literature I have differ from what is being practised in the hospital, mainly the injectable antimalarials... I had not practised as a doctor before and this was my first station...I found the changes in doses and I spent some time arguing with staff on my ward but I lost the argument because the patients got better and that is what matters really. So, my books and my knowledge seem not to be functional in these aspects... Evidence based literature would help, but I may have to struggle to convince the older people working on my ward!
>
> **Health worker 8, Iganga**

It was also noted that the periodic but inevitable revision of the national treatment guidelines contributed to the discrepancy between theory and practice, which further highlights the need for health workers' continuous updating. That takes us to another constraint identified, namely, outdated knowledge. Health workers emphasized the need to keep up to date, because they noted that some of the knowledge they had acquired previously could no longer be used. This led to the various CPD information needs discussed in Section 2.4.2. The following comment illustrates the constraint:

> During our training, we were taught to treat or prescribe medicine according to age; but now the new trend is to treat according to weight and age of a child. What I know, therefore, is no longer applicable... I keep consulting my boss — the clinical officer — but when he is not around like now he is away for a week, I get stranded because my knowledge is no longer usable
>
> **Health worker 8, Masaka**

Unavailability of medicine/facilities

Finally, some health workers reported that although they accessed information concerning medicines or treatment in general, they were not able to fully put that knowledge into practice because of a limited range of medicines available in their health units. This led to various clinical information needs (Section 2.4.2). The same applied to laboratory facilities, particularly in lower-level health units:

> I get information from various sources about drugs but I can't do much because we prescribe what is available at the health centre even though we know that another drug would be better
>
> **Health worker 5, Masaka**

> Although I know, from my initial training and from the seminars I have been attending, that management of STIs requires laboratory tests, unfortunately, we can't do some of the tests here, so we refer the patients
>
> **Health worker 3, Iganga**

Some of the constraints were moderated to enable health workers to access and use information, as the next section presents.

2.4.4 **MODERATORS**

The enhancing factors in Table 2.13a were further interpreted as moderators. The moderators were subdivided into two, namely: moderators of constraints to information access and moderators of constraints to information use. The analytical interpretation of moderators is summarized in Fig. 2.6.

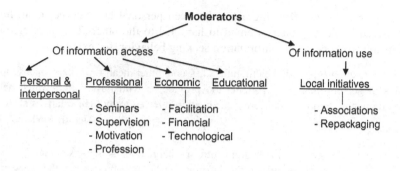

FIGURE 2.6

Diagrammatic representation of moderators.

Moderators of constraints to information access

They included, in order of importance, personal and interpersonal, professional, economic and educational moderators.

Personal moderators

The findings highlighted personal attributes such as interest and self-drive, which made health workers seek information actively. Discussing personal attributes raised several interpersonal acts, such as joining professional associations, sharing experiences and encouraging others, which were all moderators to information access. They illustrate how the value of information acts as a moderator to information access. Other aspects highlighted in this subsection were the personal but professional benefits of being heads of health units as the comments show:

> Many of us are interested in learning and updating our knowledge, so we always look for information from colleagues, DMO, etc... this enables me to know about forthcoming events such as seminars or conferences... I don't sit back and wait for information to find its way here... When I get information, I share it with others
>
> **Health worker 5, Bushenyi**

> I joined the UPMA and became an active member... I receive monthly updates and other types of information from the district branch... I have encouraged a number of people to join the association by showing them the benefits of being a member
>
> **Health worker 2, Masaka**

> As a head of a health unit, I get chances to attend meetings where various types of information is disseminated; I meet visitors and hold discussions with them and I learn certain things from these discussions. I also get chances to attend most of the seminars
>
> **Health worker 5, Iganga**

Interpersonal moderators

This included the working relations within health units, with other health workers, and patients. It also brought up the interpersonal contacts made with professional colleagues and publishers to donate sources of information. Drawing a line between personal and interpersonal, however, was quite difficult because one (personal) seemed to have led to the other. The interpersonal acts also shed light on health workers' active information seeking behavior, for example:

> Good working relationship with health unit staff and other health workers opens up various opportunities to get information... patients too provide us with information which helps us to plan as they draw our attention to different medication referred to by their relatives abroad
>
> **Health worker 2, Bushenyi**

> Having an active boss is great... the health centre in-charge attends many seminars and meetings; and every Wednesday, he disseminates the information he gathered during the week to us in a staff meeting
>
> **Health worker 8, Masaka**

> I have kept close contact with friends and colleagues in Germany where I studied... They have been very supportive... I ask them for various types of information and they provide it promptly, which supports my clinical work
>
> **Health worker 7, Iganga**

Professional moderators

The most important professional moderators were holding training sessions in rural areas, support supervision, motivation and the nature of the health profession. In Section 2.4.1, professionals were identified as an important source of information. Similarly, professional support was proven to be an important moderator to information access.

Training sessions in rural areas were ranked high as a moderator (Table 2.13a). It was noted that holding seminars in rural areas made them more accessible because it gave many rural health workers an opportunity to attend with less transport costs. Furthermore, the in-service training sessions were usually conducted using the districts' training budgets and, hence, were free of charge. The professional moderator, therefore, also moderated the financial or transport constraints.

> I have attended most of the seminars held in this area (sub county) because they are within reach... holding workshops or seminars in rural areas overcomes the transport problems and makes it easy for many rural health workers to attend
>
> **Health worker 3, Bushenyi**

Support supervision and other extension services provided by fellow health workers moderated information access. The visits by the DMO staff and other senior health workers to rural health units were reported to have solved a number of professional problems and to have answered pending questions. Such visits were highly regarded by health workers who experienced them:

> I get a weekly supervising doctor... I always look forward to his coming especially when I face difficulties... He advises me on what to do and this improves my work. I ask questions and get a lot of information from this doctor. Sometimes, I don't even wait for him to come for his routine supervision... When I have a problem, I look him up by phone and consult, but it is because he is assigned to this unit as a supervisor; so, support supervision for me is real professional support
>
> **Health worker 4, Iganga**

> other extension services for example drug inspectors and health inspectors also provide some valuable information and guidance and one can always follow up with a phone call
>
> **Health worker 3, Lira**

Motivation was repeatedly highlighted by health workers as a factor that moderates information access, whereas its absence could easily constrain information behavior:

> Motivated health workers actively seek for information and share it with others, yet the unmotivated ones hardly do so; even when they get information, they may neither read it nor pass it on to their colleagues or to lower levels... Motivation does not necessarily have to be in terms of money or allowances; it could be sponsorship to attend a seminar, refresher course or upgrading e.g. from enrolled to registered nurse
>
> **Health worker 5, Masaka**

> I have been greatly motivated after getting funding recommendations from my boss that enabled me to attend several seminars and courses where I got a lot of exposure and updating
>
> **Health worker 6, Bushenyi**

The nature of the health profession also moderated information access. For example, health workers shared their experience of gaining knowledge from the referrals they make as indicated:

> In difficult situations, we call the hospital and discuss with the seniors the cases(s) we plan to refer. Sometimes, through the discussions, we get advised on what to do and we are able to manage the cases. Secondly, when we refer patients to the district hospital, they bring back the referral forms with detailed diagnosis and treatment, which updates our standard of management
>
> **Health worker 3, Iganga**

Furthermore, the advances in the medical field and, to some extent, personal attributes did moderate access to information by compelling health workers to seek information. Such health workers were mainly the 53% of the interviewees who had made information requests to libraries in the past one year, as indicated in Section 2.4.1. Some comments:

> As a professional, I know that my field advances very fast; so to update my knowledge, I have to keep seeking for information from current literature, get some literature searches from the medical library, attend refresher courses regularly, etc
>
> **Health worker 4, Bushenyi**

> I must read to keep knowledgeable... there is no way a health professional can survive without reading because things change so fast and we keep referring to updates, books and the online databases in many situations
>
> **Health workers 1, Lira**

Economic moderators

Although the economic factors constrained information access, they moderated it in many other ways where they were mitigated. Economic moderators included transport facilitation, financial ability to access information and technological factors.

Facilitation in terms of transport moderated health workers' access to information. However, health units that had no transport identified it as a constraint.

> Most health workers at the district head office have vehicles or motor cycles, while each Government health unit in rural areas has a motor cycle and a bicycle; this helps us to move round the district easily to do our work or seek for information… Furthermore, all community development assistants have motor cycles which help them to collect and disseminate information about the health programmes they are involved in, such as provision of safe water
>
> **Health worker 5, Bushenyi**

As in other life situations, there was a general imbalance between the financial ability of interviewees or the institutions they served. The financial limitations led to the constraints presented in Section 2.4.3. However, some health workers reported that they bought books, lap tops and modems to access the Internet, and some health units did the same as indicated below.

> I have been buying basic books, newspapers, etc personally, they are expensive, but I must keep reading
>
> **Health worker 5, Bushyenyi**

> Our health unit has a small budget which we use to buy a few items such as books, newspapers, stationery and put credit on the modem for our emails and literature searches. Otherwise, we also strengthen our collection by requesting for donations
>
> **Health worker 4, Masaka**

Another economic moderator was the training that was budgeted for by the DMO to support in-service training, as already pointed out.

Technological moderators were mainly the enhanced availability of online scientific information; hence, there were some improvements in the use of online resources and the cost of computers had reduced, making it possible for institutions to improve their IT infrastructure. Nine of the interviewees who accessed scientific health information from the Internet highlighted the various benefits of the advances in technology and how the IT had moderated their information access and use:

> the new technologies have eased access to information and have changed the way we used to search for information some decades ago, which was mainly print. When I joined the MMed programme and I needed current information, I was referred to HINARI by my supervisor, and that was the first time I used HINARI and I continued using it after my MMed course… it is a great resource
>
> **Heath worker 4, Masaka**

> the word Internet remained abstract to me sometime back, but now I know how to use a computer to search for evidence based literature on the Internet and how to cite properly, which I didn't know then; I have an email account, which I didn't have, I know how to communicate scientific information using a power point presentation and a projector, which I didn't know then — these are big changes in the way I seek and use information
>
> **Health worker 8, Iganga**

> I had a phone in 2000 but it was not a smart phone, I had no lap top and I had no access to the internet, but now, I can search the internet using a smart phone, on my lap top and on office computer to access online resources, which is a major technological change that has enabled me to access information easily
>
> **Health worker 5, Bushenyi**

Furthermore, the health workers who were constrained by the slow Internet but could use an e-mail to send requests for articles/documents benefitted from the DDS provided by Makerere University Library, which moderated the constraint.

The increasing ownership and use of mobile devices greatly moderated the professional isolation experienced in the 1990s. The setting up of an electronic network by the MoH, which was highlighted in Sections 2.3.4 and 2.4.1, was another technological moderator to information access.

Media rural outreach, such as radio and television, also improved access to information as already highlighted in Section 2.4.1 (Information Sources). Further moderation of technological constraints, therefore, will make a lot of difference to the provision, access and use of scientific information by rural health workers.

Educational

With regard to educational moderators, it was noted that health workers with higher qualifications actively sought information more than those with lower qualifications. Medical doctors, for example, were active seekers, whereas most enrolled nurses and nursing assistants usually passively waited for their bosses to provide them with information, as elaborated in Section 2.4.2 (Information Behavior).

Moderators of constraints to information use

Although a quick glance would seem to indicate that there were fewer moderators than the corresponding constraints, most constraints to information use were moderated implicitly by the value of information. However, a few issues such as the unavailability of medicines and facilities in lower-level health units, which constrained information use, required economic moderators.

Local initiatives

The moderators of constraints to information use discussed here include mainly the local initiatives such as the activities of professional organizations and the repackaging of information into different formats to facilitate its access and use for those who would otherwise not do so. This further confirms what was presented in Section 2.3.4, that local initiatives were taking control of the identified constraints to information access and use.

Activities of professional organizations such as UPMA repeatedly appeared in the private midwives' interviews, whereas UMA was highlighted by medical doctors. The associations moderated constraints to information access by providing the much needed information such as reports, newsletters and other documents to their members, as well as running seminars and conferences for them (as indicated in Section 2.4.1). Furthermore, they were also moderators of constraints to information use because they provided information that was relevant and applicable, with the required details and level. This was exemplified by interviewees who reported (under constraints) that some international journals were generally not applicable to Uganda's rural situation, although they were good for updating. In contrast, the Uganda Medical Journal produced by UMA was considered very relevant as the comments show:

> I find UMA activities very informative... there is always something to learn... I am able to update my knowledge and to apply the new knowledge in what I do, which improves patient management and health care in general because we pass on this information to others...UMA runs scientific conferences which provide current and relevant information from recent researches and other developments in the field
>
> **Health worker 1, Lira**

With regard to repackaging of information in print and electronic formats, health workers pointed out that the locally produced documents such as the Uganda Health Information Digest and CPD materials were very relevant and applicable because they focused on Uganda's health needs (digest) and took into account the situation in Uganda's health units (CPD). The services provided by Makerere University Library and the Uganda Health Information Network (UHIN) implemented by the Uganda Chartered Healthnet were the main local interventions that were greatly appreciated by the interviewees. Commenting about the Digest and CPD materials, for example, some interviewees pointed out that:

> The digest is very relevant to our needs, and we have applied the knowledge in various ways, unlike some texts from international sources which are too technical or inapplicable to our rural situation
>
> **Health worker 3, Masaka**

> It is good to see that the oldest University in the country has produced something for us in rural areas, so that we can benefit from Ugandan expertise... the topics focussed on are the common health problems in Uganda, which make it relevant and applicable
>
> **Health worker 6, Iganga**

The repackaging of information in smaller documents such as pamphlets and newsletters made it easy for busy health workers to read. This is well supported by the findings (Table 2.9) of health workers who overwhelmingly reported that smaller pieces of information were the easiest ways to deliver information to them, and the use of text messages sent on mobile phones proved to be a very practical and fast way to access information.

Furthermore, the repackaging of information in audio-visual formats such as radio and television programs that were presented by senior medical professionals were considered important because they updated health workers' knowledge. Local interventions, therefore, moderated information use in various ways.

2.5 DISCUSSION OF INFORMATION ACTIVITIES

Sections 2.3 and 2.4 are drawn together and discussed to wrap up the information activities.

2.5.1 INFORMATION SOURCES

Various characteristics of information sources were identified. The most important ones were accessibility, interactivity and reliability or credibility. For both women and health workers, accessibility of an information source was an important factor. Women explained, for example, that even though health workers were an authentic source of information, they were generally not accessible; hence, neighbors, friends or TBAs were preferred as a source by some women.

Health workers generally identified colleagues as an important source because they were within reach. This was supported by studies such as Farmer and Richardson (1997), who noted that "reliance on informal sources of information is as a result of their convenient access...nurses appear to select information on the basis of accessibility rather than on quality" (p. 98). Other studies that reported reliance and/or preference of colleagues by health workers because of their accessibility include those by Tumwikirize et al. (2009) and Younger (2010). It was also noted that in extreme situations, lack of an easily accessible source could bring an information process to a halt.

Similarly, both women and health workers emphasized the issue of interactivity. Radio, for example, which was ranked highly as a source of information was also criticized by women for what they described as a "one-way channel." In contrast, seminars/workshops and other training sessions were perceived by health workers as the best channel because of interactivity, among other things. Previous studies such as those by Johnson and Meischke (1991) indicated the importance of interpersonal interactions as effective sources of information and as being well suited to handle individual needs and questions because of the interactivity that was referred to as "immediate feedback." Paek et al. (2008) also identified interaction and interpersonal communication as important factors affecting health education and behavior. The ability to interact with information sources has received attention from information retrieval researchers and a model based on the interaction of texts and users was reported by Vakkari (1999). He pointed out that "this interactionistic approach supposes that information searching is inherently an interactive process between human beings and texts intermediated by an IR system" (p. 823).

This book has shown that the interactionistic approach need not be limited to texts intermediated by an information retrieval system. It can and has been demonstrated to be between human beings (eg, in seminars/workshops) or between human beings (interviewees) and sources such as texts but intermediated by human beings (eg, seminar tutors or senior professionals).

Despite some shortcomings (eg, lack of interactivity), radio emerged first as an actual source of information for women and was highly rated by lower-level health workers in rural areas. In the case of women, radio was followed by health workers and seminars, whereas for health workers it depended on the level and location, such that doctors in urban and peri-urban areas had their main source of scientific health information as online resources/databases followed by printed sources, seminars/workshops, and CPD sessions. However, doctors in rural areas and lower-level health workers mainly depended on printed sources followed by seminars/workshops and other CPD sessions, seniors and colleagues, and then radio. Some rural health workers, however, benefitted indirectly from online resources through the DDS provided by Makerere University Library.

The reasons for radio ranking high may be due to its increased access at household and rural area levels (due to the FM stations). Lack of, or limited access to, other sources of information as highlighted under constraints could also have made some interviewees more dependent on the radio. This is consistent with what Wilson (1997: 565) observed: "for authoritative information, . . .people consult more than one source of information partly as a result of dissatisfaction with the information they receive from one source or another. Thus, if the information dissemination activities of organisations in the health field are inadequate, the main fall-back source may well be the media."

It has also been reported in mass media literature that the greater the distance from a referent, physical or social, the more important the role of the media, and that interpersonal communication, formal and informal, tends to compete with the media in such areas (Severin and Tankard, 1988). This was indeed true in this book as interactivity or interpersonal communication in seminars/workshops, with colleagues in the case of health workers, or with friends and relatives in the case of women combined to compete favorably with radio and other media as actual sources of information.

In choosing or selecting what program to listen to on the radio or watch on TV, the perceptions of program format, quality and professional content were strongly related to the choice health workers made. In contrast, women were generally less selective; only 25% reported having chosen particular health programs.

A combination of factors, namely, low educational levels, lack of libraries in the rural areas studied, and limited availability of printed sources in local languages tended to encourage an oral information culture among the women. That culture supports a high degree of word-of-mouth communication. Furthermore, the factors that encourage an oral information environment could have contributed to audio sources scoring highest as actual source of information compared to printed sources such as books.

Heath workers generally pointed out that the best channel to deliver information to them was not cheap, and that it was mainly print materials and training such as seminars/workshops or refresher courses, whereas the easiest way was by mobile phones and printed sources. However, women identified radio followed by health workers as the best ways/channels to deliver health information to them. The easiest channels were LCs, mobile phones, faith-based events/leaders, women's groups and the use of different channels, which were generally similar to those identified by the health workers. However, there were differences in the perceptions of women and health workers with regard to the best sources/channels of information for the women. For example, health workers generally maintained that the best channel/source of information for women was health workers, and yet the limited access to health workers in rural areas had raised great concerns and made women prefer other nonprofessional sources because they were accessible. Health workers, therefore, talked about an ideal situation that, according to the women, had not been realized.

Health workers then recommended a multichannel package, or as many avenues as possible, to be adopted to ensure that women accessed health information in rural Uganda. They also indicated that radio was an important channel in their health education work, and that people who listen to the radio participated actively in the discussions. Furthermore, some health workers observed that the audio-visual aids such as video/films and posters only facilitated the dissemination activities of health workers but would not replace or take over their role; hence, health workers have to find a way of fulfilling their health education role in rural areas.

Sources such as the LCs were also of great importance to women, but not to health workers. However, religious institutions or leaders facilitated access to health information in rural areas, and health workers commended them for indirectly providing information by sponsoring seminars/training and providing printed and audio-visual materials. This was particularly true for health workers in church-aided units. Furthermore, women appreciated religious institutions/leaders as a direct source of information through counseling and preaching, and indirectly by organizing seminars and talks. In addition, health workers and women pointed out that NGOs, both local and international, contributed significantly to the provision of information in rural areas mainly through seminars, workshops and other training programs, as well as printed and electronic information.

Illiteracy is among the factors commonly reported in Sub-Saharan African studies as hindering access to information. Women interviewed, however, reported that availability of appropriate printed materials was more of a problem than illiteracy. They pointed out that illiteracy does not totally stop rural illiterate women from accessing information in printed form if that information was repackaged to suit their level (simplified and translated in the vernacular) and made accessible to them either directly or indirectly through LCs, women councils/groups and/or school children. They highlighted several advantages of printed information; for example, once information is available, it can be kept for reference, and even those who were not around when it was disseminated would get it in full without being distorted or misreported like oral information (Musoke, 2007).

It was noted that books and pamphlets came last as actual sources of information used by women (Table 2.1). This could be due to the fact that there were generally no libraries in the rural areas studied. Some urban areas, namely Lira and Masaka towns, had public libraries (resource center), but they had rarely been used due to a number of factors. Of the forty-eight women interviewed, only two (from Lira) reported having used the public library (for books and/or newspapers); however, two (from Masaka) reported having used some primary school book collections. The four women had a secondary-level education. Hence, there appeared to be a link between education level and library or book use. Generally, it was noted that the role of libraries and other formal information sources (except radio) in the provision of health information to the rural women interviewed was negligible. Several studies from Sub-Saharan Africa highlight the challenges of libraries as sources of information (Mooko, 2005; Adjah, 2005, Gadau and Lwoga, 2013).

Furthermore, apart from the medical doctors and some clinical and nursing officers, the rest of the professional health workers interviewed did not have access to formal libraries; but they depended on small collections within the health units. Hospital collections were proven to be important to staff working within hospitals and were also a source of reference for lower-level health units' staff supervised by the hospitals. Sizeable libraries were in academic institutions, such as Makerere University Main and Medical Library in Kampala district, which had provided various information services to health units through information literacy training and DDS. Other academic libraries in Mbarara University of Science and Technology in the west had served some interviewees, whereas Gulu University in the north focused on its primary clientele and had not extended services outside the University, which was understandable given the demands and the available human and information resources.

Some health workers, who were unaware of how to exploit the resources at the Makerere University Main and Medical Library without physically going there, appreciated the information sharing during the interview when the author informed them about the DDS service and provided a brief demonstration on how they could order and receive the documents by e-mail or post. This would greatly ease access to the needed literature, particularly where the Internet was slow.

Despite the shortcomings highlighted regarding access to printed sources, they were still perceived as the most important source of information for rural health workers, mainly because of the information technology challenges in rural Uganda. This led to recommendations such as those made by Dr Bewes, a retired surgeon: "computers and internet have their value at the centre of knowledge disposal, but it was not an either-or question. If the electronic is not appropriate, it need not detain us; but if it is, let's have more if it.... We should not lose the hard worn values of print... Just as the TV did not kill the radio or the cinema, and none of them has yet killed the book, I think that the CDs and the internet will simply be added to the many tools of the information seeking humans."

That brings the discussion to the IT dilemma. Electronic sources, such as the Internet, were reported by women and some health workers only as potential sources of health information. Of the thirty-four health workers interviewed, for example, only nine had accessed online databases, including HINARI and other electronic sources such as CDs, which they found very useful. They added that the major change in the information environment by 2015 was the reduction in reliance on printed sources as a result of enhanced access to online scientific information. There was a "digital divide" among the interviewees, with the majority reporting the Internet and related technologies as a challenge that had denied them access to online resources, hence their preference for printed sources. However, all the health workers reported having accessed information from the MoH electronic network that sent text messages, about various health issues, to mobile phones of all registered health workers in Uganda. The increase in ownership and use of mobile phones was an important channel of speedy information exchange among the health workers and the women, which led to overcoming the communication gap and isolation experienced in the 1990s. All women and health workers interviewed had mobile phones. Six of the health workers (doctors) pointed out that they used their smartphones to access the Internet and to search for basic and scientific information. It was further noted that the Internet mobile data were becoming less expensive due to competition among service providers, thus making it possible to search the Internet longer than before. With the exception of the TBAs and nursing assistants, all health workers interviewed had an e-mail address, but only two women had. Some health workers, particularly doctors, used e-mails to request articles from Makerere University Library, thereby indirectly accessing electronic resources.

Furthermore, the HMIS was an important source of electronic data and information on Uganda. Similarly, the District Population database provided information on population and related issues.

As reported in Chapter One, there is Internet connectivity in Sub-Saharan African capital cities, but most rural areas had challenges. In Malawi, for example, by 2015, the fiber optic network was being installed from the coast, but the costs were still very high. Bandwidth challenges were regional, but land-locked countries such as Malawi and Uganda faced higher costs. The National Research and Education Networks (NRENs) played an important role in this area, but they targeted research and educational institutions. The IT infrastructure has therefore remained a fundamental access factor in the region.

Related to the challenges of IT infrastructure was the low computer literacy among rural and/or older health workers. Rhine and Kanyengo (2000) observed that development partners seem to concentrate on access factors and neglect the training of individuals who utilize the IT. In the past, many medical and paramedical training schools did not provide adequate training in informatics, which left the graduates, at that time, computer-illiterate. Some of those graduates became policy makers who would not advocate for a facility they were not familiar with, when the tight budgets had to cater for other seemingly pressing needs. Several health workers who graduated before 2000, however, reported the need for computer literacy sessions in addition to the health information literacy skills.

In addition, studies of women's use of IT have indicated, for example, that women were slower in using the Internet even though their interest was high, and that some of the differences in the patterns of usage could be gender-related, such as attitude toward technology, heavy workload leaving limited time for leisure and other factors. Marcella (2002) recommended that Internet sites aimed at women could provide an alternative space for communication that women might find useful. However, Olatokun (2008) pointed out that unless gender issues in the use of IT are addressed in Africa, women, particularly those in rural areas, will continue to be excluded from the IT scene.

2.5.2 INFORMATION NEEDS

The literature on user studies indicated that information needs (and active seeking) had persistently attracted the attention of information researchers. This could have a historical explanation in that the social world has been transformed over a period of time. Such transformation results from the changing lifestyles, technology, new diseases and other environmental factors. Consequently, information needs and behavior have tended to change because of:

- advances in medical knowledge and practice in the case of health workers;
- a growing population that keeps worsening the health worker/patient ratio;
- new diseases such as AIDS that demand care from the family and, hence, increase the burden on women as care providers;
- developments such as new technologies, for example, the Internet, which has brought about information overload and the challenges of quality, cost and access.

Although no previous study seems to have been conducted on this subject in rural Uganda, such changes in the social world make information needs and behavior evolve and develop. This may explain why needs and behavior will continue to attract information researchers in future.

The study on which this book is based found out that women's needs and their information behavior were quite dynamic; even within a single incident, they could change from latent to active needs, and hence, from passive behavior to active seeking. That finding, therefore, differs from studies that take information needs as static, such as those reported by Vakkari (1999: 823): "…the model has treated information needs as prefixed and static."

Some studies concentrated on the met information needs. Soto (1992: 305), for example, reported in her summary of findings that "academic staff tend to identify information with printed literature and they perceive information needs as those satisfied by that information source."

In this book, however, unmet needs for information came to the fore front at the pilot study stage and the interview schedule was modified to include a section on that issue. The findings have been important in highlighting that only 26% of the health workers interviewed had their information needs satisfied, whereas 30% were partially satisfied, and the rest (44%) were not satisfied. Furthermore, the findings have also highlighted the link between information needs and constraints, because the unmet or unsatisfied needs for both women and health workers became constraints to information use.

It was noted that health workers' information needs were mainly cognitive (clinical, training, preventive care, causes), but some of their personal or professional needs, such as further education, were affective because they were geared toward achieving a higher qualification or getting a promotion. However, women's critical and active needs were generally affective because of the emotional effect of life-threatening diseases or critical health problems. The latent needs were generally cognitive (treatment, causes, detection, prevention). However, some latent needs such as overcoming constraints and misconceptions were affective. As already indicated, women's needs were dynamic; therefore, as they changed from latent to active, they also changed from cognitive to affective and led to active or purposive information seeking.

Women's selflessness, as primary and most times sole care givers in their nuclear and extended families, was highlighted in the study when their personal health information needs usually remained latent, except in critical situations. This confirms what Oxaal and Baden (1996) observed: "the value women place on their own health can be influenced by a number of factors such as informational barriers and low self-esteem... Women may not regard their own pain and discomfort as worthy of complaint until it is so debilitating..." (p. 19).

Both women and health workers had various needs for information. The needs have clearly been demonstrated as well as the demand for information. Confirming that demand, health workers pointed out that when they go to, for example, immunization programs in rural areas, women insist that the sessions should start with health education. Furthermore, health workers repeatedly highlighted the need for continuous professional development to keep abreast of the advances in their profession and indicated how outdated knowledge had constrained their activities. The study findings differ from some claims that there was low demand for information from African health workers and that they do not use information resources even when they are accessible. The findings presented in this book differ from those claims in several ways: the demand for information was demonstrated in critical incidents and in the way health workers actively sought information despite the various obstacles. The need to keep up to date was equally highlighted as an information need. Furthermore, the use and attribution of value to information are presented in Chapter Three.

Finally, the findings have confirmed further the importance of providing timely information. Some health workers lamented about the need for information on Ebola before the outbreak in Uganda and neighboring Congo, which they had not been able to get. Several women also pointed out that they asked some health workers who informed them that because research on Ebola was still ongoing, there was hardly any information to disseminate yet. If the available information on prevention had been disseminated when those needs were identified and after the Ebola problem was reported in neighboring countries, then Ebola might not have claimed so many lives in Uganda in October and November 2000. Fortunately, Uganda learned from the previous experience and was able to provide excellent information to both the health workers and the public, which facilitated the prevention of Ebola in 2014. Furthermore, using the experience gained, Uganda was able to provide medical personnel support to other African countries. The unmet health information needs remain a challenge to the health of human beings.

2.5.3 **CONSTRAINTS**

The major constraints identified were socioeconomic. The economic constraints affected access to information sources and underpinned other factors such as IT infrastructure, staffing of health units and transport. The social factors played equally a big role in constraining information access and use. The women's findings clearly elaborate how gender, culture and language factors constrained their information access and use. Health workers were more constrained by professional factors, such as staffing, and, to a lesser extent, by social factors, such as nepotism and security (Musoke, 2007).

The findings have highlighted constraints to both information access and use, and have attempted to bridge a gap in information studies research that was reported by Wilson (1997): "there is scope for a great deal of work on how the wide range of possible intervening variables actually affects specific search situations or particular groups of users" (p. 570).

For both women and health workers, the quality of information accessed constrained its use. Health workers pointed out that some of the information they accessed was irrelevant, inadequate, questionable or repetitive, and that there was an information overload on the Internet that was time-consuming, whereas lower-level health workers raised the issue of some information being too technical. Women, however, pointed out that unclear and incomplete information constrained them from using it.

The constraints to information use also highlighted the more personal nature of the use of information. Some health workers, for example, talked more specifically about their personal experiences concerning information use, although in a few instances they mentioned the public. This was in contrast to the section on constraints to information access, in which health workers talked about their juniors being passive and demoralized. To illustrate that point further, when doctors referred to some newsletters as elementary and to certain project reports as lacking the details that would benefit them, they did not mention that they "used" those documents in another way by discussing their contents with the junior staff they supervise! During one interview, for example, the researcher observed that one doctor had put a "For ward 2a" sticker on a newsletter; when asked why she was sending it to the ward rather than the hospital library, she replied that her juniors read it first and they discuss the contents with her before sending it to the library. Hence, although the newsletter's content was inadequate to the doctor's needs, it could have been adequate to her juniors. The fact that those "uses" were not mentioned and only the constraints were reported emphasizes the personal nature of information use.

That points to the need for information providers to strike a balance between the appropriate quality of information so that those who need advanced information get it and those who need the elementary or simple type also do, but not at the expense of the other. Both quantity and quality of information constrained information access and use, and should be addressed in information provision strategies.

Furthermore, personal factors such as attitude, perceptions and emotions were reported to have constrained women's access to and use of information. Similarly, the educational level of health workers appeared to have affected their information seeking behavior such that the more educated a health worker was, the more she/he sought information and from more evidence-based sources.

These findings are consistent with those of Wilson (1997), who pointed out that the "Characteristics of an information source may constitute a barrier, either to information seeking behavior or to information processing, and that personal variables may be either psychological or demographic" (p. 568). Furthermore, Savolainen (2015) pointed out that the impact of cognitive barriers is mainly negative because they block, limit or hamper information seeking or give rise to negative reaction such as frustration.

In Section 2.3, women pointed out that some health workers' negative behavior had halted their information seeking acts and made them prefer other more approachable, although unprofessional, sources. This generally constrained women's access to information from health workers. It was interesting to note that some health workers raised the same issue, which confirmed what had been reported by women. They pointed out, for example, that if a health worker insults a woman, she may never go back to consult her/him, and she will pass on information about that mistreatment to her friends and relatives. In a few instances where it was reported, the attitude and negative behavior of health workers acted as obstacles in the information environment, thereby hindering women's access to information.

The World Health Organization (2009) observed that "although women are the main users of health care, insufficient attention is given to their needs and perspectives ... and to the constraints that they face in protecting their health or in accessing or making the most of available services. Key concerns for women seeking health care include respect, trust, privacy and confidentiality — values that are often compromised in busy facilities, particularly among certain age and social groups" (p. 75).

Given these concerns, the Uganda MoH introduced a toll-free "8200" information service in 2013 to enable patients to report problems they encountered in health units (which is a moderator to information access). The women who knew about the facility pointed out that it had improved service delivery and given confidence to the consumers, thereby increasing health care seeking in Government health units and the use of health facilities, which all contribute to improved health outcomes.

Women also identified "*limited access to professional health workers in rural areas*" as one of the factors that hindered their access to health information. Some health workers pointed out that shortage of staff leading to a heavy workload and compounded by transport challenges made it difficult for them to reach all the rural areas under their supervision regularly to hold health education sessions. Health workers also identified transport as a major factor affecting rural women's access to health units. However, the availability of mobile phones had greatly enhanced the consultation on urgent issues either directly or indirectly through the village health team members.

Commenting about the shortage of health workers in rural areas, the DMOs in the four research districts admitted that it was difficult to reach all rural areas because of the shortage of staff. This was caused by the failure to get the established posts of health workers filled. Health workers tend to prefer working in towns or in private rather than Government health units. The author confirmed, during the selection of the health workers' sample, that some categories of health workers were difficult to find deep in rural areas.

It was also noted, with concern, that the existence of drug shops/pharmacies in villages, some of which were run by untrained personnel, put the rural communities at risk. Among other things, the scarcity of trained health workers in some rural areas and the fact that licensed health units were few and/or far away made rural people seek medical advice from nonprofessional drug shop owners because they were accessible. Some health workers lamented that the media advertisements of some medicines and their doses seem to have legitimized further the selling and buying of such medicines without prescriptions! This had several consequences, for example, underdosage, overdosage and mixing or taking different brands of medicines for the same illness as reported by some health workers who had attended to patients in critical conditions after mixing antimalarial drugs. These issues raised the concerns of women and health workers, as well as the MoH; for the women, "*limited access to professional health workers in rural areas*" remained a major factor that constrained their access to professional advice. The MoH increased the monitoring of operations of drug shops/pharmacies in the country to address the problem.

When the responses from women and health workers about factors affecting women's access to health information were compared, they were found to have a lot in common. However, there were differences in emphasis that seem to have arisen from the different experiences and/or perceptions. For example, women identified language as a problem on its own, whereas most health workers referred to it as an educational factor. If one considers English language alone, then it becomes an educational factor; however, some women in Iganga district reported having received materials in Runyankole (language spoken in Bushenyi), which were not easily understood. Furthermore, some health workers pointed out that, without an interpreter, they fail to conduct health education in the community because they do not speak the local language. Hence, the language factor goes beyond education. It is Uganda's national challenge of having many languages. The educated people are better off because they can access the materials/sources in English. However, even in countries such as Kenya, which have a national language, it was recommended that the rural women's health information service and the training should be conducted in local languages and the information materials should be translated (Muthoni and Miller, 2010).

Furthermore, health workers did not report late invitations to seminars and workshops as a negating factor, probably because they did not realize that this was one of the reasons why some women did not attend those events. That issue, however, was not raised by women in Bushenyi and Lira districts, but mainly by those in subcounties approximately thirty kilometers or more away from Masaka and Iganga district headquarters. Distance and poor network to receive text messages on phones, therefore, seem to have affected the receipt of invitations by some interviewees.

One of the assumptions the study had at the beginning was that distance from Kampala and from the district head office to rural areas affected access to information. The findings from both women and health workers confirmed that distance was a constraint. It affected access to information directly, as highlighted in the section about geographical/spatial constraints, and sources such as libraries; access to information was also indirectly affected by technological constraints in rural areas. If the financial and technological constraints could be addressed, then distance or physical location would not be too much of a problem.

Studies conducted in developed countries, however, also identified distance as a problem for community health workers. Soto (1992), for example, pointed out that "Distance between the site of work and the library is a more serious problem for community service dentists who work in clinics around Sheffield ... Distance is also a deterrent for using the library in the case of general dental practitioners" (p. 312).

As pointed out in the information sources section, enhanced access to electronic information was enjoyed where the IT infrastructure was supportive, but mainly by doctors and a few clinical and nursing officers. Women reported the Internet as a potential rather than actual source of information due to various constraints. Other women lamented their failure to call radio presenters to ask questions because the telephone lines were busy. However, even when the telephone lines were not busy, such as in the developed countries, some groups of people, especially the passive audiences, were reportedly not utilizing the service. For example, Belderson (1999) observed that "Although the use of telephone information service has been advocated by some investigators... these are only likely to be of use to those who are already motivated - the majority simply do not seek information in such a purposeful way... Indeed, although such a service does exist in Sheffield (Sheffield Health Line), no interviewee mentioned being aware of it or using it for nutrition information" (p. 235).

It was noted that women generally identified the immediate or basic factors that constrained them from accessing and using information. The sociocultural factors or constraints seem to have affected women greatly and to have influenced their values, needs and practices. Women tended to conceptualize those basic constraints not in isolation, but in terms of their relationships and interactions with the family and community. This closely relates to what Omaswa and Crisp (2014) indicated: "Ill health, pregnancy related problems and disease in Africa are all made far worse by poverty and wider social issues such as conflict, poor education and social division, some of which are the direct legacy of colonialism" (p. 38).

As already noted, discussing constraints highlighted some overlap with information sources. The findings also revealed a strong linkage between constraints and information needs as presented below. In some cases, constraints triggered needs, whereas the reverse was true for others.

Constraints that led to information needs

Constraints to information use led to information needs, for example, when husbands refused to let women use contraceptives (Section 2.3.3), this triggered a need for information. Knowing the value of information, some women reported that they decided to learn more about FP advantages to be able to convince their husbands to let them use contraceptives.

This was also true in the case of health workers. When some health workers realized that they could not put their knowledge into practice (Changes in Medical Practice, Section 2.4.3), they needed to know why, for example, the antimalarial drug recommended doses in textbooks differed from those practically administered on the ward.

Information needs that led to constraints

In other situations, unmet information needs led to constraints to information use. The women's data highlight various interdependencies and relationships between unmet information needs, sources of information and constraints to information use, for example, regarding the immunization program. Although all women interviewees reported that they accessed information about immunization, according to some women the information they had accessed left some unanswered questions. For example, some women needed information about the safety of the vaccine; when that information need remained unsatisfied, women decided not to have their children immunized; therefore, they were constrained to use the information they accessed about immunization dates, venue and age of children to be immunized. There were a number of issues to note:

i. The dynamic nature of information needs that changed from passive to active. In this case, women accessed information about polio immunization passively; afterwards, they needed to know about the safety of the vaccine, which became an active need.

ii. At the time of the interview, the unmet information needs had led to constraints to information use, but a lack of information about the safety of the immunization program was both an unmet need and a constraint to information use, as illustrated in A and B, Fig. 2.7.

The unmet information needs led to constraints to information use. If women obtained the information/clarification about, for example, the safety of the vaccine (to satisfy their information needs), then the constraints could be overcome and the information could be used by taking

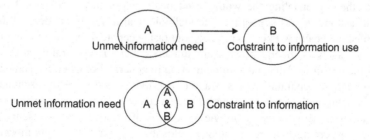

FIGURE 2.7

The effect of unmet information needs on information use.

Musoke, M.G.N., 2001. Health Information Access and Use in Rural Uganda: An Interaction-Value Model. PhD. Thesis, University of Sheffield; Musoke, M.G.N., 2007. Information behaviour of primary health care providers in rural Uganda: an Interaction-Value model.

J. Doc. 63 (3), 299–322.

children for immunizations. Hence, the more information needs that are met or satisfied, other considerations being held constant, the fewer constraints to information use and, therefore, the more successful the preventive measures, such as polio immunization or FP, would become.

iii. It was further observed that although there had been massive campaigns for polio immunization and only rumors or word-of-mouth messages about the alleged relationship between polio immunization and the origins of HIV/AIDS, some members of the public seem to have responded more to the latter by resisting participation in the immunization exercise. Hence, a few informal but negative messages from friends or relatives tended to have more impact than the massive positive messages from formal and authoritative sources.

It was also noted that misconceptions about immunization and AIDS stemmed from both local and international sources. A German physician, Geisler (1996), for example, reported that "97% of the persons, who have HIV, were purposely infected with the virus . . . and the majority are in Africa. . . Only roughly 3% of infected persons were exposed to HIV unintentionally, either by sexual intercourse or as new born infants by their mothers." Similarly, Hooper (2000), in his document "River of tears," linked the origin and spread of HIV in Africa to immunization: "I propose the origin for HIV-1 Group M, is pertaining to Chat, an experimental oral polio vaccine (OPV). . . But for an AIDS epidemic or pandemic to occur, I believe what is needed is a mass exposure, as through a vaccination campaign."

In Uganda, it was noted that WHO, UNICEF, Rotary International and others, the major supporters of the Uganda National Immunization Days (NIDs), had not only approved the safety of the immunization program but also supplied the anti-polio vaccine and the equipment (including sterilizes) used to dispense it during that period. It was also noted that the Uganda Government put forward a reward to anybody who would provide scientific evidence to prove that the polio vaccine was contaminated or unsafe as it had been alleged, but nobody was able to do so! However, some mothers pointed out that they were harassed by husbands for immunizing their kids, whereas others were reluctant to have their kids immunized. Hence, unmet information needs remained a strong constraint to information use.

2.5.4 **MODERATORS**

Informal or social networks and the value of information were the major moderating factors identified. Through interpersonal interactions and the repackaging of information by the MoH, professional associations, librarians and other information providers, various constraints to information access and use were moderated, thereby enabling the women and health workers to access and use information. Where the IT infrastructure was supportive, the technological advances moderated the various constraints and enabled the health workers to get easy and fast access to scientific information. Other technological advances such as mobile phones did moderate various constraints such as professional isolation among the health workers and communication in general. For example, some women pointed out that they were able to communicate easily and to get invitations to meetings/seminars by text messages, thereby overcoming the late invitations and distance constraints experienced in the past. Putting together issues about moderators inevitably yields sources of information because there was moderation of constraints concerning sources, and it also yields the constraints that were moderated.

Leaders (local, women and religious) were the major moderators of constraints to information access. In addition to providing information directly, women reported that religious institutions and leaders moderated information access in various ways; for example, they organized and funded drama or films, invited health workers to give talks or seminars, and collaborated with LCs and Government. This was very important because women pointed out that their husbands did not object to their attending faith-based activities. Hence, besides moderating constraints concerning sources, for example, access to health talks/seminars, the church and other religious institutions also moderated the gender constraints to information access. Furthermore, when religious institutions organized free film and drama shows, they moderated the economic constraint to information access. Women had reported that some people failed to attend such shows because of entrance fees.

The church also moderated information access by playing a role in shaping people's beliefs, attitudes and values. This was commended because it influenced people's positive behavioral change and promoted health. However, it became controversial when AIDS prevention and FP issues were discussed. Some health workers pointed out that the practice of mobilizing people to use contraceptives contradicted the church's stand, which constrained information use and, consequently, acceptance of the practices. That issue, however, was raised only by health workers. Women reported religious leaders, beliefs and practices as moderators of constraints to information access without highlighting the contradictory messages as an issue.

In line with that experience, Czarnecki (2015), drawing from a study conducted in the United States, observed that devout Catholic women had two contradictory cultural schemas, religion and secular, which provided them with different cultural resources to resist various activities experienced by other nondevout Catholic women, and they drew on religion to find value and meaning in their lives. In that context, therefore, the church remains a moderator rather than a constraint.

Another moderator was the local council system (LCs). The LCs were commended by both the women and health workers as sources of information and moderators of constraints to information access when they invited health workers and organized drama and film shows. The LCs were unanimously reported by the women as the easiest way to provide information to them because of its ability to reduce the identified constraints to information access. In a few situations and areas, however, the positive role of LCs was reported to have been reduced by some LC executive members who rarely organized health education and promotion sessions and related information events.

Women's experiences, negative and positive, coupled with the value of information did moderate a number of constraints through interactions with their social networks. Indeed, women's negative life experiences played a big role in their information activities, but specifically in the moderation of constraints. The women with disabilities, for example, reported that they mobilized and convinced people to take children for polio immunization, using themselves as examples that no mother would wish to see her child to become. In other cases, women who had many and poorly spaced children because they did not learn about FP used that experience to sensitize their family members and other people, and they managed to convince them to use FP methods.

Similarly, the positive experience of women who learned about FP methods and successfully applied them was used to convince others who had misconceptions. Women pointed out that they act as role models when sensitizing others and showing them the benefits of contraceptives. Some women confirmed that they were able to overcome various misconceptions after friends and/or relatives shared their positive experiences and advised them on what to do. Hence, women's experiences greatly moderated misconceptions.

In the case of health workers, interpersonal interactions, professional activities and local initiatives emerged as the key moderators to information access and use. The advances in technology also moderated constraints to information access for the health workers whose IT infrastructure was supportive, as already indicated.

Finally, it was noted that the repackaging of information by the MoH, professional associations and librarians (specifically the Uganda Health Information Digest) in the case of health workers, by women's groups and leaders in the case of women, and by various other organizations was a major moderator of constraints to information access and use. The other moderating factor was the value of information itself, as presented in Chapter Three.

2.6 CONCLUSION

This chapter has provided the methodology used in the study on which the book is based. It then summarized and presented women's experiences, perceptions, views and information behavior that were recorded during the interviews, analyzed, and interpreted by the author. The dynamic nature of women's information behavior was demonstrated. Furthermore, women vividly expressed their needs for information, the preferred sources, what constrained them to access and use information, and how some of the constraints had been moderated.

Similarly, the chapter summarized the evidence from the health workers' personal and professional experiences, as well as the perceptions of their information environment. As professionals in the ever-growing field of medicine, health workers' information needs were generally more pronounced and led to active information behavior, which was different from that of the women. However, the individual models, which were inductively derived from the women and health workers' data, had similar core and main categories. This made it possible to generate an overall model that is presented in Chapter Four.

Finally, both women and health workers demonstrated the value of information and how it had moderated the constraints to facilitate information access and use. The next chapter, therefore, focuses on the value attributed to information by its users, and this is one of the main issues that this book set out to present.

REFERENCES

Adjah, O.A., 2005. The information needs of female adult literacy learners in Accra. Inf. Dev. 21 (3), 182–192.

Baker, L.M., 1995. A new method for studying patients' information needs and information seeking patterns. In: Lloyd-Williams, M. (Ed.), Health Information Management Research, Proceedings of the 1st International Symposium. University of Sheffield, Department of Information Studies, Centre for Health Information Management Research, Sheffield, pp. 67–75. April 5–7.

Bantebya-Kyomuhendo, G., 1997. Treatment Seeking Behaviour Among Poor Urban Women in Kampala, Uganda. PhD. Thesis, University of Hull.

Belderson, P., 1999. Food Choice in Older Adults: the Role of Nutrition Information. PhD. Thesis, University of Sheffield.

Brettle, A., 2015. Measuring the impact of health libraries in practice. EAHIL conference. https://eahil2015. wordpress.com/workshop-presentations (accessed 30.07.15).

Bryant, S.L., Gray, A., 2006. Demonstrating the positive impact of information support on patient care in primary care: a rapid literature review. Health Inf. Libr. 23 (2), 118–125.

Carter, I., 2002. Facilitating change: the role of animators. Development in practice 13 (1), 83–89.

Conte, C., 2015. Crossroads: women coming of age in today's Uganda. www.amazon.com/crossroads-women-coming-Todays-Uganda/dp (accessed 10.07.15).

Czarnecki, D., 2015. Moral women, immoral technologies: how devout women negotiate gender, religion and assisted reproductive technologies. Gend. Soc. 29, 716–742.

Dominick, J.R., 1996. The Dynamics of Mass Communication. fifth ed. McGraw-Hill, New York, NY.

Ellis, D., 1993. Modelling the information seeking patterns of academic researchers: a Grounded theory approach. Libr. Q. 63 (4), 469–486.

Farmer, J., Richardson, A., 1997. Information for trained nurses in remote areas: do electronically networked resources provide an answer? Health Libr. Rev. 14, 97–103.

Gadau, L.N., Lwoga, E.T., 2013. Information seeking behaviour of physicians in Tanzania. Inf. Dev. J. 29 (2), 172–182.

Geisler, W., 1996. AIDS: origin, spread and healing. AIDS Anal. Afr. 6 (3), 4.

Glaser, B., 1978. Theoretical Sensitivity. Sociology Press, Mill Valley, CA.

Glaser, B.G., Strauss, A., 1967. The Discovery of Grounded Theory: Strategies for Qualitative Research. Aldine de Gruyter, New York, NY.

Hooper, E., 2000. The river of tears. In: AIDS Origins. http://www.aidsorigins.com/.

Hughes-Hassell, S., Agosto, D.E., 2007. Modelling the everyday life information needs of urban teenagers. Youth Inf. Seeking Behav. II, 27–61.

Johnson, J., Meischke, H., 1991. Women's preferences for cancer information from specific communication channels. Am. Behav. Sci. 34, 742–755.

Kaleeba, N., Ray, S., Willmore, B., 1991. We Miss You All: AIDS in the Family. Women and AIDS Support Network, Harare.

Marcella, R., 2002. Women on the web: a critical appraisal of a sample reflecting the range and content of women's site on the internet, with particular reference to the support of women's interaction and participation. J. Doc. 58 (1), 79–103.

Marcus, R., 1993. Gender and HIV/AIDS in Sub-Saharan Africa: the cases of Uganda and Malawi. BRIDGE Report 13. Institute of Development Studies (IDS), Brighton.

McQuail, D., 1994. Mass Communication Theory: An Introduction. third ed. Sage, London.

Momodu, M., 2002. Information needs and information seeking behaviour of rural dwellers in Nigeria: a case study of Ekpoma in Esan West local Government area of Edo State, Nigeria. Libr. Rev. 51 (8), 406–410.

Mooko, N.P., 2005. The information behaviour of rural women in Botswana. Libr. Inf. Sci. Res. 27 (1), 115–127.

Musoke, M.G.N., 2001. Health Information Access and Use in Rural Uganda: An Interaction-Value Model. PhD. Thesis, University of Sheffield.

Musoke, M.G.N., 2007. Information behaviour of primary health care providers in rural Uganda: an Interaction-Value model. J. Doc. 63 (3), 299–322.

Muthoni, A., Miller, A.N., 2010. An exploration of rural and urban Kenyan women's knowledge and attitudes regarding breast cancer and breast cancer early detection measures. Health Care Women Int. 31, 801–816.

Ndira, S., et al., 2014. Tackling malaria village by village: a report on a concerted information intervention by medical students and the community in Mifumi, Eastern Uganda. Afr. Health Sci. 14 (4), 882–888.

Nuijten, M., 1992. Local organisation as organising practices: rethinking rural institutions. In: Long, N., Long, A. (Eds.), Battlefields of Knowledge: The Interlocking of Theory and Practice in Social Research and Development. Routledge, London.

Odini, C., 1995. A Comparative Study of the Information Seeking and Communication Behaviour of the Kenya Railways and British Rail Engineers in the Work Situation. PhD. Thesis, University of Sheffield.

Olatokun, W.M., 2008. Gender and national ICT policy in Africa: issues, strategies and policy options. Inf. Dev. 24 (1), 53–65.

Omaswa, F., Crisp, N., 2014. African Health Leaders: Making Change & Claiming the Future. Oxford University Press, Oxford.

Oxaal, Z., Baden, S., 1996. Challenges to Women's Reproductive Health: Maternal Mortality. BRIDGE Report 38. Institute of Development Studies (IDS), Brighton.

Paek, H., et al., 2008. The contextual effects of gender norms, communication and social capital on family planning behaviors in Uganda: a multilevel approach. Health Educ. Behav. 35 (4), 461, http://heb.sagepub.com/cgi/content/abstract/35/4/461.

Patton, M.Q., 2002. Qualitative Evaluation and Research Methods. Sage Publications, London.

Palsdottir, A., 2007. Patterns of information seeking behaviour: the relationship between purposive information seeking and information encountering. In: Bath, P., et al., (Eds.), Proceedings of the 12th International Symposium for Health Information Management Research (ISHMR). University of Sheffield, Sheffield, pp. 3–15.

Phillips, S., Zorn, M., 1994. Assessing consumer health information needs in a community hospital. Bulletin of the Medical Library Association 82, 288–293.

Rhine, L., Kanyengo, C., 2000. The development and use of the Guide to Medical Resources website at the University of Zambia medical library. Proceedings of the 8th International Congress on Medical Librarianship, July 2–5. Library Association, London. <http://www.icml.org/Tuesday/>.

Savolainen, R., 2015. Cognitive barriers to information seeking: a conceptual analysis. J. Inf. Sci. May. http://jis.sagepub.com/content/early/2015/05/27 (accessed 30.05.15).

Severin, W., Tankard, J.W., 1988. Communication Theories: Origins, Methods, Uses. second ed. Longman, New York, NY.

Soto, S., 1992. Information in Dentistry: Patterns of Communication and Use. PhD. Thesis, University of Sheffield.

Sorrentino, R.M., et al., 1990. Personality functioning and change: Informational and affective influences on cognitive, moral and social development. In: Higgn, E.T., Sorrentino, R.M. (Eds.), Handbook of Motivation and Cognition: Foundations of Social Behavior, vol 2. Guildford Press, New York, NY, pp. 193–228.

Standing, H., Kisekka, M., 1989. Sexual Behaviour in Sub-Saharan Africa: A Review and Annotated Bibliography. Overseas Development Administration, London.

Strauss, A., Corbin, J., 1990. Basics of Qualitative Research: Grounded Theory Procedures and Techniques. Sage Publications, Newbury Park, CA.

Strauss, A., Corbin, J., 1998. Basics of Qualitative Research: Techniques and Procedures for Developing Grounded Theory. second ed. Sage, London.

Tipping, G., Segall, M., 1995. Health Care Seeking Behaviour in Developing Countries: An Annotated Bibliography and Literature Review. IDS Development Bibliography 12. Institute of Development Studies, Brighton.

Tumwikirize, W.A., et al., 2009. Access and use of medicines information sources by physicians in public hospitals in Uganda: a cross-sectional survey. Afr. Health Sci. 8 (4), 220–226.

Turner, B., 1981. Some practical aspects of qualitative data analysis: one way of organising the cognitive process associated with the generation of Grounded theory. Qual. Quant. 15, 225–247.

Uganda National Academy of Sciences, 2014. Mindset shifts for ownership of our continent's development agenda. Report of the committee on ensuring country ownership of Africa development agenda in the Post-2015 Era.

Urquhart, C., et al., 2003. Critical incident technique and explication interviewing in studies of information behaviour. Libr. Inf. Sci. Res. 25 (1), 63–88.

Vakkari, P., 1999. Task complexity, problem structure and information actions: integrating studies on information seeking and retrieval. Inf. Process. Manage. 35, 819–837.

Vogelsang, R., Oltersdorf, U., 1995. Information needs and preferred information sources on nutrition topics: results of a qualitative study. Appetite 24.

Wallman, S., 1996. Kampala Women Getting by. James Currey, London.

Were, M.K., 2014. Community health, community workers and community governance. In: Omaswa, F., Crisp, N. (Eds.), African Health Leaders: Making Change & Claiming the Future. Oxford University Press, Oxford, pp. 109–124.

Wilson, T., 1981. On user studies and information needs. J. Doc. 37 (1), 3–15.

Wilson, T., 1997. Information behaviour: an interdisciplinary perspective. Inf. Process. Manage. 33 (4), 551–572.

WHO, 2006. Building foundations for eHealth: progress of member states. Report of the WHO Global Observatory for eHealth.

World Health Organisation, 2009. Women and Health: Today's Evidence Tomorrow's Agenda. WHO Press, Geneva.

Younger, P., 2010. Internet based information seeking behaviour amongst doctors and nurses: a short review of the literature. Health Inf. Libr. J. 27 (1), 2–10.

THE VALUE OF INFORMATION AND EFFECT ON HEALTH OUTCOMES

3

3.1 VALUE OF INFORMATION CORE CATEGORY

As indicated in Chapter Two the fifth preliminary category that emerged from data analysis was value of information. It was the value attributed to information and the subsequent actions reported by the interviewees. Although the study on which the book is based was user-centered, it did not set out to measure the value of information. The value, as presented in the book, emerged inductively from the qualitative data. Interviewees reported that when they accessed information and used it, some of that information changed their states of knowledge, values, beliefs and behavior. This led to the various actions that put the knowledge acquired into practice or applied the information gained in various ways. The interviewees indicated that the application of the knowledge gained solved various health problems and met health information needs, which improved and promoted health. Various examples of application of knowledge are presented in the subsequent sections of the chapter.

The ability of information to solve problems fulfills the second delimiting analytical rule of the Grounded theory concerning a core category (Glaser, 1978), as highlighted in Chapter Four.

It was further noted that the interviewees, as users of information, judged the information they accessed and attributed (or did not attribute) value to it. The difference that information made to people was, therefore, conceptualized as the value of information.

The subsequent sections of the chapter present, among other things, examples of the value of information as attributed by its users, and its effect and impact on health care.

3.2 USE OF INFORMATION AND ATTRIBUTION OF VALUE BY WOMEN

This section presents the value of information and the actions that resulted from it.

The raw data, from the responses to the question "what was the information used for?", were first coded to get the concepts and then compared with each other to eliminate obvious overlaps when possible.

Women reported that the information they had accessed was used in the prevention and treatment of diseases, to know the causes of illness, to improve health, to stay healthy, to make decisions, to make choices, to overcome constraints and misconceptions, to cope with illnesses, to support the community/self-help, to change behavior, to change attitude, to participate in information dissemination/awareness raising, to identify information sources and for general health

Informed and Healthy. DOI: http://dx.doi.org/10.1016/B978-0-12-804290-8.00003-8

knowledge. All the women interviewed indicated that most of the information they used was also shared with others; therefore, participation in information dissemination was ranked first, followed by prevention of diseases and then treatment or management of illness was third. It was also noted that active information behavior was exhibited in most of the information activities reported in this section.

3.2.1 VALUE OF INFORMATION

As indicated in Section 3.1, the value of information refers to what the information was perceived to mean, its role and significance in women's personal and family lives, and in their activities as leaders. Most of the concepts that emerged from the analysis of information use were similar to those that emerged from information needs in Chapter Two (Section 2.3.2). This is consistent with what Wilson (1981: 5) observed: "information use ought to point most directly to the needs experienced by people."

The value of information had the following subcategories:

Causes and prevention, treatment, health knowledge, improved health, overcome misconceptions and attitudinal change, overcome constraints and coping. Other concepts that emerged from the analysis of information use were "constraints to information use," which is in Chapter Two (Section 2.3.3), and the rest are presented as actions in the next section (Section 3.2.2). The women's value of information is summarized in Fig. 3.1.

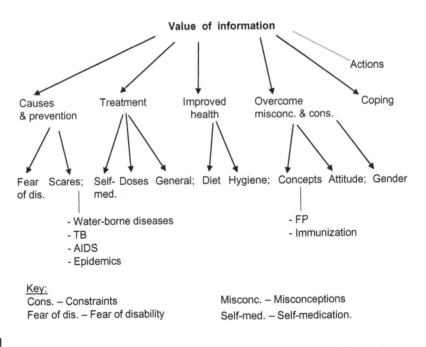

FIGURE 3.1

An illustration of women's value of information category.

Causes and prevention

Information was valuable in the prevention of illnesses by knowing their causes. Knowledge of how diseases are caused and transmitted was reported to have led to the control of disease vectors and health promotion in general as the comments show:

> I learned that malaria is transmitted by mosquitoes, so I do everything possible to keep this home free from bushes and stagnant water where mosquitoes breed, and to close windows and doors before dark... Since I started doing all that, my household members take long to suffer from malaria; actually, we may spend a year without an attack, yet in the past, it was a frequent problem in this home
>
> **Woman 2, Bushenyi**

There were also some common concepts that relate to prevention, for example, fear of a disabled child, financial benefits and warning bells or being scared as presented below:

Fear of a disabled, sickly or abnormal child

Concerns were mainly about the failure to immunize children who would consequently suffer from immunizable diseases, and how such experiences made people value the information they received and had their children immunized to avoid the problems as illustrated in the comments:

> Having a crippled child is not a joke... the messages on TV and on posters which show children playing football while a crippled child is looking on, are very touching and moving... I got to know that once children are not immunised, they won't be protected from diseases such as polio which have no cure or from killer diseases such as tetanus... I have used that information effectively by taking my children for immunisation and I finish the recommended doses
>
> **Woman 4, Lira**

> When I heard about polio, I was very interested in the topic because I am lame ... I listened attentively and concluded that my problem could be due to the fact that I was not immunised. I then hurried to take my children for immunisation so that they don't suffer the way I did. I have also encouraged friends and other people to do the same, giving myself as a live example; this has convinced many to have their children immunised
>
> **Woman 2, Iganga**

In such situations, the quality of information received and the fear or the experience of disability enabled people to use the information about immunization. Such people were different from those whose information needs about the safety of the vaccine remained unmet, which constrained them from using the information, as reported in Chapter Two (Section 2.3.3, Constraints).

Information that was used to prevent diseases was also reported to have had some financial benefits, for example:

> I learned about the importance of immunisation and my younger children got immunised. Since then, they have been a lot healthier than their older siblings... This reduces our medical expenses because they do not suffer from diseases such as measles which used to disturb our older children
>
> **Woman 5, Iganga**

> where families have been vigilant in implementing what we learn from health education sessions, the expenditure on health at household level has reduced than it was fifteen years ago, and yet we had smaller families then ...Although doctors charge more now than they did then, the number of times one consults a doctor has greatly reduced because of avoiding most of the diseases that we can prevent
>
> **Woman 4, Masaka**

Scares or warning bells

The information received about the prevention of diseases/health problems such as AIDS, waterborne diseases and epidemics in general acted as warning or alarm bells. The quality of information and the fear of death facilitated its use as the comments show:

> The AIDS drama I watched was very depressing ... and I even cried...I also get disturbed to see how AIDS patients go through the physical and psychological pain and trauma, leaving their loved ones especially children who really suffer after the death of their parents... This scares me and I keep reminding my husband to avoid extra marital sex otherwise we would die!
>
> **Woman 4, Iganga**

> The message 'always drink boiled water' has made my family develop a habit of boiling drinking water to the extent that I would rather not drink water if it is not boiled ... because if you escape worms and other simple water borne diseases, you would die of cholera ... so, why drink unsafe water at all!
>
> **Woman 12, Masaka**

> It is better to prevent cholera because one may never get a chance to be cured ... cholera patients die in a very short time...We were very fortunate because we got very good, simplified and timely information about cholera; that is why it didn't reach this area (Kagango sub county) - everybody, young and old, is aware and vigilant.
>
> **Woman 1, Bushenyi**

Treatment

Interviewees also reported that the information they received facilitated the treatment of diseases and management of patients. Treatment included self-medication, doses of medicines and general management (treatment choices are in the next section: Actions—Making Decisions); for example:

> I learned how to make ORS (oral rehydration salts) which I find handy in a rural setting where health units are far ... in case of diarrhea, I no longer have to go to a health unit for ORS
>
> **Woman 11, Lira**

After getting information about antimalarial drugs, for example, and the importance of completing the full dose, several women reported that they had noticed a positive change when the guidelines were followed because malaria was properly cured; for example:

> We used to take aspirins for malaria but now I know that antimalarials such as chloroquine are the ones that treat malaria ... We were also not sure about the doses but after the seminar, I know that medicine is administered according to age: babies, children and adults; and that for

effective results, a full dose has to be taken ... there is a tendency of stopping medicine as soon as one feels better ...but I make sure that my family members complete the prescribed dose, and now I see that malaria attacks are no longer frequent.

Woman 7, Masaka

Management of illnesses, health problems and patients in general was also supported by information. Women reported that they were able to manage their health problems as well as those of their children and other family members using the knowledge gained. Being able to manage such problems gave women confidence and had various social benefits to the family as indicated in the comments:

I got to know that for hypertensive people, it is important not to overwork, get enough rest and follow the guidelines about diet e.g. avoid fatty foods, alcohol ... This has stabilised my pressure. Similarly, with ulcers, now I know that it is important to feed properly and to reduce worries ... This has helped us a lot as a family because when I am healthy I spend my time doing productive work for the family

Woman 4, Masaka

My children have a skin problem; so, the seminar I attended recently was very relevant to me. I learned a lot, asked questions and got some good seminar notes. I know that I am not supposed to apply any cream on the affected skin without the advice of a medical personnel, and that we should not use any soap for bathing... This has greatly improved my children's skin problems ... in fact, one can't tell easily that they have a problem

Woman 6, Masaka

In some situations, the critical incidents were a blessing in disguise because they were reported to have made women gain valuable knowledge and skills, for example:

My four year old son drunk paraffin accidentally... I needed some first aid knowledge urgently... One of my neighbours is a scout; I called him and he administered the first aid and the boy started vomiting... Then I took him to a clinic for further management. However, I made sure that I learn the first aid... I went back to my neighbour when I was relaxed, and practised it. So, this incident opened my eyes and helped me to learn something very useful

Woman 10, Masaka

Health knowledge and improved health

Information was valuable for improving health and enabling women to stay healthy as well as to maintaining the health of their family members. This mainly focused on diet and hygiene. More than 30% of women interviewees reported how information had made them knowledgeable about diet/nutrition using locally available foods, which greatly improved the health of their families, especially children, for example:

I learned what food value(s), e.g. proteins, carbohydrates, calcium, vitamins and minerals, are found in the local foods and what each does to the body or what its absence could result into. I also learned about the growth of children, care and feeding e.g. solid food starts at 6 months, the dangers of early weaning, and how proper feeding protects children from preventable diseases. I asked questions in the seminar and understood clearly what I didn't know e.g. the importance of a high protein diet for children, and the value of millet porridge. After the seminar, I put my

> 7 year old son on the recommended diet for about six months now … the boy, who was a weakling, has now gained weight and strength and several people have already asked me what I give him because the difference is so obvious, yet I feed him on simple things which we all have here in the villages
>
> **Woman 1, Iganga**

> The information has enabled me to maintain healthy teeth and I encourage my family members to do the same… regular brushing is important, and we limit the sweet things our children eat.
>
> **Woman 1, Lira**

Generally, information enabled people to gain knowledge on various health topics, which was a source of pride and confidence because they were able to share with others from an informed base and applied the knowledge in various ways.

Overcome misconceptions

It was noted that the quality of information received and the interpersonal moderators played a significant role in overcoming misconceptions. Information was valuable in disproving concepts and/or overcoming negative beliefs and attitudes about family planning (FP), immunization and other medical practices. When such beliefs were overcome, women were able to use contraceptives, take children for immunizations and/or use medical instead of traditional practices/services. Hence, "overcoming misconceptions" triggered various decisions. Some decisions were made as a direct result of overcoming misconceptions, whereas others were made as a result of the value of information in general and are presented in the next section (Section 3.2.2, Making Decisions).

Furthermore, there were various examples of attitudinal change as a result of information use that brought about some professional, social and other benefits, such as:

> The seminar I attended about disability enabled me to understand the causes of disability …which can be explained medically. Since then, I developed a different view about disability, and this has helped me to handle my pupils with disability better
>
> **Woman 2, Masaka**

> When I attended a seminar, I learned about the importance of child spacing … It was not easy to convince my husband but after some time, he accepted and I started FP. By controlling birth, our love became stronger because we now have a lot more time together without getting interruptions from a baby crying or pregnancy problems. This helps the marriage a lot. In addition, since I am not pregnant almost every year as I used to be, I have more time and strength to do my community work as LC4, because my area of operation and supervision is quite big. These are real benefits of FP - stable and happy marriage; healthy children; strong, healthy and more productive (not reproductive) mother; good planning for the future, which I use to sensitise the women I lead to join FP
>
> **Woman 4, Iganga**

These comments are consistent with the Steps to Behavior Change (SBC)[1] model, which posits that behavioral change among individuals and groups occurs in five stages: knowledge, approval,

---

[1]SBC was set-up for the design and evaluation of family planning communication projects.

intention, practice and advocacy. People usually do not take any action (especially with regard to something new such as condom use, immunization and FP) unless they have sufficient knowledge of it, have developed a positive attitude toward it, and have talked with others about it. The more of these things they have done, the greater the likelihood that they will take action.

Overcome constraints

Information was used to overcome various constraints, such as gender. Some women indicated that the advice they received enabled them to convince their partners, for example, to use contraceptives. This was valuable given the number of women who reported this problem in Chapter Two (Sections 2.3.2 and 2.3.3, Information Needs and Constraints, respectively). It was noted that some misconceptions led to constraints and vice versa, which were overcome by the value of information in some situations.

Coping

Information was valuable for enabling individuals and families to cope with the various diseases and emotions related to the disease or health problem. Several interviewees indicated that information helped to alleviate both the psychological and physical suffering:

> I use the information I get to take care of myself and specifically to reduce getting sickle cell disease attacks e.g. I was advised to always take the medicine provided by the sickle cell clinic, keep warm and to have a balanced diet as much as possible... Whenever I learn something that can keep me healthy, it gives me hope and courage to face the next day
>
> **Woman 5, Lira**

> My husband died of AIDS recently... when I was still nursing him, I wanted to know how to manage the situation and how to handle my other family members to accept... I also wanted to know whether I am safe or not ... I read pamphlets and books about HIV/AIDS... I listened to radio programmes, attended seminars and health education sessions about AIDS and learned quite a lot which greatly helped me to handle the difficult and stressful situation... Anyway, when he died, I continued getting information until I gathered courage and went for a blood test ...So, in such situations, information is very valuable because it enables one to do some essential things like in my case, to counsel family members to accept the problem and for me to take a blood test and to know what to do in order to prolong life ...The information I get also helps me to stand firm; since the problem has come, being ignorant about the disease won't take it away, in fact it would make the situation worse
>
> **Woman 2, Masaka**

3.2.2 ACTIONS

The section presents activities, behaviors or processes that were triggered by the value of information; in other words, the value of information acted as a stimulus to the actions reported. Some of the actions were a consequence or a by-product of information use. Further interpretation of data from the women showed that actions had the following subcategories, which are illustrated in Fig. 3.2:

- Participation in information delivery or dissemination and awareness raising
- Making decisions
- Community support

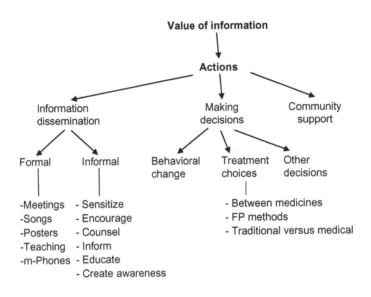

FIGURE 3.2

Analytical interpretation of women's actions.

Participation in information dissemination and awareness raising

All the women interviewees reported that after accessing information and using it, they then shared it with other people, thereby participating in information dissemination and awareness raising. The finding agrees with Wilson (1981: 5), who noted that information "use may satisfy or fail to satisfy the need and, in either event, may also be recognised as being of potential relevance to the needs of another person and, consequently, may be 'transferred' to such a person."

As community leaders, women interviewees were commended for providing a scarce commodity, information, in rural areas. It is a commodity in the sense that information seemed to gain value as it passed along the information production chain, which agrees with one of Braman's (1989) conceptualizations of information. The findings also agree with the two-step flow model of communication (Defleur and Dennis, 1989) in that information moved from the source to women leaders, who then passed it on to others whom they inevitably influenced.

It was also noted that the various information dissemination activities moderated constraints to information access and use. The examples given in this section were taken from a large set of similar quotes, which illustrate the many ways in which interviewees disseminated information or performed awareness raising. The activities were subdivided into formal and informal methods of information delivery.

Formal

The formal activities included: meetings and feedback sessions at Local Councils (LCs), in church and other places; drama, songs and poems; posters and teaching of pupils in schools and other people. Messages sent by mobile phones could be either formal or informal. The activities were held within the community to ensure that information reached as many people as possible. Some examples of meetings/health education or feedback sessions are given below:

As a women's representative of the village Local Council (LC1), I pass on the information I get e.g. from the seminars I attend or from radio, to the community members during the bimonthly village meetings…We discuss, they ask questions, and in some cases, I have to follow them up house-to-house to check whether they have put in practice what we discussed; for example, removal of mosquito breeding sites such as stagnant water

Woman 1, Bushenyi[2]

Every month, we conduct one day seminars at the parish level to sensitise the community about different health topics using the information we (members of the LC3 task force) get from various sources. Whenever I have urgent information to disseminate, I request the priest to announce it after the church service. Furthermore, as the chairperson of women LC3, I also hold meetings of women councils twice a year for two days each and they are funded by the LC3 (sub county) budget. When they go back, women councillors disseminate the information further

Woman 3, Iganga

I learned about the advantages of spacing children - to the child, the mother and the family; when I shared this information with members of my (women's) group, they found it very useful and decided that we invite someone from the FP office to give a similar talk here in the village for the benefit of women and other members of our community… and if possible, we should continue holding such sessions periodically

Woman 3, Lira

Information was also disseminated through songs, plays and poems, for example:

I watched an interesting play about AIDS and I got some phrases from it, which I used to write a short poem about AIDS and culture, titled, 'lako' (widow inheritance) which has been distributed to other areas.

Woman 3, Lira

We collect information to assist us in composing songs and drama about the control of common diseases such as malaria, which affect our community

Woman 11, Lira

Others used text messages on mobile phones and handmade posters to disseminate information:

As a secretary for information on this village, I have to find a way of keeping people informed; for example, I send 'sms' regularly by phone, but I also use manila sheets to put up messages and/or notices e.g. invitation for meetings where we disseminate information fully. For example, during the cholera epidemic, and with the help of health workers, we held several health education sessions and used posters to inform people about the causes and prevention of cholera.

---

[2]That interviewee was the LC1 secretary for women in one of the villages researched in Bushenyi district. Several interviewees from Bushenyi (Women 2–4) kept referring to the meetings she organized because they provided information about various development issues, including health.

> Similarly, during the National Immunisation Days, we mobilise parents in the same way to take children for immunisation …Using manila and markers, I write the dates and venue where immunisation will take place and put the posters in public places …I check on them regularly to see whether they are removed so that I can replace them, but usually they don't get torn off
>
> **Woman 3, Masaka**

Besides using posters, drama, mobile phones and meetings to inform people, some women disseminated information in their professional activities such as teaching and held demonstrations for their women's groups as illustrated by the comments here:

> I use the information I get about AIDS and other topics in teaching my pupils at school as well as members of our women's group …I also use the information to do some counselling
>
> **Woman 9, Lira**

> I pass on the information I get to my pupils and I also tell them to share with their parents; for example, basic hygiene — washing hands, drinking of boiled water; the importance of immunisation and prevention of malaria
>
> **Woman 2, Masaka**

> Since I learned how to extract milk from soya, I hold talks and demonstrations for other women to learn, and I have followed up to see the implementation which is very good
>
> **Woman 1, Lira**

Informal

Although some women did not disseminate information formally, they all reported to have disseminated information on various health topics informally to immediate family members and other relatives, neighbors, friends and the community where they lived and/or worked. Women used the informal means to encourage, inform, sensitize, counsel, educate and create awareness, as indicated in the comments. Such means, although informal, demonstrate the active processes of information dissemination.

> I got information about FP quite late after I had stopped child birth; so, I didn't use that information personally, but my daughters have greatly benefited from it … I cite my bad experience of producing many and poorly spaced children and I strongly encourage them to use FP methods… They all have well-spaced children after taking my advice
>
> **Woman 7, Iganga**

Furthermore, the nature and quality of information received facilitated its use and in some situations moderated constraints, for instance, parents who reported that they were uncomfortable to discuss sex and related issues with their children until they attended concerts as illustrated in the comments:

> I have learned some 'straight talk'[3] … I talk to my adolescent children these days about the causes and dangers of 'slim'
>
> **Woman 5, Bushenyi**

---

[3]"Straight talk" was a program on radio, TV and newspapers in Uganda that sensitized parents to talk to their children about sex and related topics such as condom use, HIV/AIDS and pregnancy.

> The concerts by school children are very good for some of us shy parents because they give us a starting point... When I attended the recent one in which my children participated, it had songs about AIDS and a play about adolescent problems such as pregnancy and drugs, which helped me to start talking to my children about these issues...Since then, I discuss with them from time to time; otherwise, I had no courage to start this important discussion with them
>
> **Woman 8, Masaka**

As presented in Chapter Two (Section 2.3.4, Moderators), the provision of informal health information moderated various constraints. Other authors such as Haythornthwaite (1996) described how informal communication networks are formed and reformed as the needs and the environment change, resulting in a constantly emerging network. Thus, informal information exchange routes develop based on local needs. The ability to locate these routes and the ability to define the groups and roles that arise to respond to the needs are important to a study of access to health information in an African rural setting. This is because the current proliferation of new ways of accessing information in the developed world using the Internet or even the basic information sources such as libraries and other information units are still problematic in many parts of rural Sub-Saharan Africa.

It was also noted that although women mainly accessed information passively, they passed it on actively in their social networks. Women in this book can, therefore, be described as passive information recipients but active disseminators at the same time. Hence, passivity was generally limited to the initial stages of their information process, which again highlights the dynamic nature of women's information behavior. In social networks or informal communication, information is made useful and its value grows by being shared with and/or forwarded to others.

Making decisions

The information women accessed and used empowered them to make decisions about what to do or avoid, where to go and which treatment to take. Making decisions included behavioral change, treatment choices and other decisions. This agrees with what Gray (1991: 281) indicated: "our behaviour is governed by our knowledge. We respond not so much to physical reality as to our understanding of it... the processes by which we understand the world and use that understanding to guide our actions."

Behavioral change

Some women indicated that the information they had received made them take various decisions, for example, to change behavior to prevent diseases such as AIDS. Fig. 3.3 illustrates the activities, behaviors and/or processes triggered by the value of information as follows:

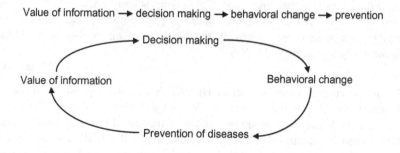

FIGURE 3.3

The value of information and resultant actions.

The comments made by women who decided to change their behavior to prevent HIV/AIDS include:

> Information about AIDS transmission and prevention has helped me personally... I have to guard against 'sleeping around' anyhow... I have to be faithful to my partner hoping that he will do the same
>
> **Woman 1, Lira**

> The information I receive from different sources keep reminding me about the presence of AIDS ... To avoid it, I tell myself that every man has it except my husband... So, I have to do serious 'zero grazing' with one partner, my husband, and forget all about the 'side lines' but whether my husband does the same, I don't know! I talk to him about AIDS and he sounds convinced... extra marital relations have resulted in AIDS and eventual death... anyway, for me I am doing my best
>
> **Woman 8, Iganga**

Women living with HIV/AIDS also shared their experiences of how information enabled them to make decisions concerning sexual relations because this could prolong their lives when they prevent other infections, for example:

> When I got to know about the risks of re-infection, I decided to stop... I am not getting any partner because I don't want to shorten my life. This has kept me going since my husband died of AIDS two years ago
>
> **Woman 5, Bushenyi**

> I have known about the dangers of having multiple partners, unsafe sex, etc; to protect my life, I have to limit my 'social movements', actually to stop getting men so that I can live a little longer for the sake of my children. I also stopped using the same razor blades, etc with my family members
>
> **Woman 4, Bushenyi**

Some women also reported to have made decisions to change habits such as alcoholism, which, to some extent, could expose one to AIDS. The changes had various social and economic benefits as indicated:

> After learning about the dangers of over boozing, I gave up going to bars ...This helps me to save time and money which I used to waste in bars, and I feel stronger and healthier than when I was still drinking ... It was a big problem, for the children, having both of us alcoholic. Since I quit drinking, I have tried to show my husband the positive aspects; he stops briefly, but lapses back ... So, I have a task to make him stop ... Drinking is such a bad habit that it can lead to many other temptations or risks, for example, since I stopped drinking, I also reduced the temptations of having extra marital sex ... actually staying in bars long exposes one to risks such as rape which can easily infect one with AIDS
>
> **Woman 4, Masaka**

This book has shown that awareness about HIV/AIDS was the highest of all the health topics (see chapter: Access and Use of Information by Women and Health Workers, Table 2.5, information that was accessed). If that high awareness could be translated into behavioral change, then the spread of AIDS would be greatly checked in Uganda. Mundell (2000) reported that for the first time in Africa since the beginning of the AIDS epidemic, one country—Uganda—was experiencing a decline in the new HIV infections among its adult population. This brought hope to other countries that epidemics can be controlled by effective information use and behavioral interventions.

The Ugandan findings, however, differ from studies such as those cited by Ginman (2000), which typify a gap between wealth of information and desired behavior in developed countries. One of the examples cited was women's knowledge about cancer and their actual behavior and lifestyles. Furthermore, the "inconsistency between wealth of information and behaviour was common in people's attitudes towards food compared to their eating habits... An intensive information campaign does not always seem to lead to the desired knowledge or the desired behaviour" (p. 12).

The Uganda national AIDS awareness program was, therefore, commended for the decline in the infection rates as a result of information access and use, which led to positive behavioral change. Many players were involved, for example, the Government of Uganda, NGOs, faith-based leaders, the media and other information professionals.

Treatment choices

Information enabled women to make choices between the available medicines, different FP methods, traditional and medical treatment, etc. It was noted that by the time women made a choice between the different FP methods, for example, they had overcome the constraints and misconceptions about FP that were reported in Chapter Two.

Furthermore, women made choices between traditional and medical treatment. The choices were influenced by the information received and sometimes by other people's experiences, for example:

> I used to treat myself and even my children with traditional herbs first, but now I know that this is risky ... when my children get malaria, for example, I rush them straight to the health centre...I no longer give them herbs first because by the time I reach the health unit, it may be too late, my neighbour's baby died like that!
>
> **Woman 3, Bushenyi**

> When my first born was 8 years, she developed a swelling in the abdomen. Her teacher advised me to take her to a district referral hospital. When I reached there, I was told that the girl was going to have a surgical operation. However, when my husband learned of the operation, he said we should take the girl to traditional practitioners. I asked the health workers to give me more information about the girl's problem so that I could educate my husband about the disease and its possible causes. The information I got helped me to convince him that the problem will be best handled by medical rather than traditional methods. The girl was operated and recovered very well
>
> **Woman 7, Iganga**

Other health decisions

As already indicated, decisions such as starting FP and taking children for immunizations were a direct consequence of overcoming misconceptions and/or constraints. The decisions were influenced by a number of factors, for example, the involvement of national dignitaries, the fear of disability as highlighted in Section 3.2.1 (Causes and Prevention) and the quality of information received that enabled people to overcome doubts and misconceptions and, in turn, met their information needs about the safety of the vaccine and/or the effectiveness of immunization, as expressed below:

> I really had doubts about the polio vaccine until I watched a video where I saw 'Maama Janet' (the President's wife), the Vice President, the Minister of Health and other Government officials taking their children or relatives for polio immunisation... Then I realised that the exercise was safe... This dispelled the rumours about the safety of the polio vaccine and made us take our children for immunisation, as well as mobilising others
>
> **Woman 3, Iganga**

Similarly, women indicated that they accessed information that enabled them to overcome misconceptions and made decisions about FP as indicated in the comment:

> For some time, I had resisted using contraceptives because of the stories I heard, and that is why my children are poorly spaced... I heard that after using pills, for example, when I decide to get a baby, I could get an abnormal one and that when I stop taking pills, I could repeat my periods several times a month and could get cancer and other complications... One of my friends directed me to an FP clinic where I enquired about all those issues and I was told they were not true. So, I started taking pills for two years now and I am fine ... nothing has happened to me ... in fact my periods are better since I started using pills and my children are going to be properly spaced
>
> **Woman 2, Iganga**

In addition, information enabled women to make decisions involving, for example, breastfeeding children for longer periods, being tested for HIV and using condoms:

> The need to breast feed for a long time was emphasised in a seminar I attended... I used to breast feed for about 6 months, but when I learned that the longer the better, I decided to breast feed my youngest for over a year, and I have seen the benefits because she is healthier than the rest
>
> **Woman 11, Masaka**

> I used to worry very much about my HIV status because my partner had other women... When the sub county health centre got HIV testing facilities, we were informed and encouraged to make use of the facilities and we also had pre-test counselling. This was very useful information for me. I gathered courage and decided to get my first test which was negative. When the results turned negative, I decided to leave that partner because I don't want to die... Since then, I have had five other tests ... all negative which really gives me a peace of mind
>
> **Woman 9, Iganga**

Community support and self-help

The value of information also stimulated support to the community health needs. Examples of needs are: rehabilitation of water sources, identifying the sick in the community and encouraging them to go to health units, mobilizing parents to take children for immunizations and applying the various things women learned. Some comments:

> I know the importance of safe water ... together with my neighbours, we have mobilised resources and protected a spring nearby which we use; we also collaborate with a water source committee to clean the well every 2 weeks
>
> **Woman 3, Lira**

> I identify pregnant women in the community and talk to them about the risks of delivering in villages and hence the importance of using health units, and attending ante-natal clinics during pregnancy. People appreciate this advice and some come back for further information
>
> **Woman 7, Masaka**

> I have learned a lot of things from seminars and other sources which I am trying to implement so that our area's health can improve. As a leader, I must give a good example; so, I try out these things first … Recently, I learned how to make an improved cooking area which is less smoky to protect women's lives who spend a lot of time in kitchens… I have mobilised women to have such a facility and I follow up to see the implementation, which is greatly appreciated
>
> **Woman 3, Iganga**

The next section presents the experiences of health workers as information users.

3.3 USE OF INFORMATION AND ATTRIBUTION OF VALUE BY HEALTH WORKERS

It was noted that health workers used information generally in *updating* their knowledge and skills, and subsequently in:

  i. *preventive care* (health education and other preventive activities);
 ii. *clinical work* (diagnosis, treatment and delivery of health care);
 iii. *academic work* (training, project proposals and production of documents);
 iv. *professional support* (guidance and support to colleagues and juniors);
  v. *administration* (of health units and activities);
 vi. *detection* (of diseases/health problems);
vii. *personal health.*

These findings relate closely to those of Brettle (2015), who investigated the use of health-related evidence summaries by medical practitioners and reported that 29% read the evidence summaries for continuing professional education, 34% read to answer a specific question or address a specific issue in practice, 24% read for general interest and 13% read for other unspecified reasons.

The health workers' uses of information, which emerged from the data inductively, are discussed further in the next two subsections 3.3.1 and 3.3.2.

3.3.1 VALUE OF INFORMATION

The value of information referred to the role and significance of information in health workers' professional and personal lives, and in their activities. It is the value attributed to information that emerged from the data, as highlighted in Section 3.1.

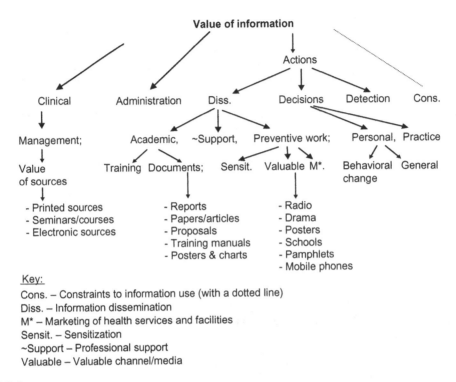

FIGURE 3.4

An illustration of health workers' value of information category and its subdivisions.

Generally, the concepts that emerged from the analysis of information use (eg, clinical, preventive and academic) were similar to those that emerged from information needs but with some differences in the details. This shows that information was valuable in satisfying (some of) the needs of health workers. Furthermore, "updating" was found in all the subcategories of value of information, just like in information needs; it was found to be a means to the end, rather than an end in itself; for example:

> I get to know about the new trends and methods of ante- and post-natal care, and the management of labour which enables me to do my midwifery work better
>
> **Health worker 2, Masaka**

> New malaria drug combination/regimes which are much superior than the old ones are suggested for resistant cases... this updates our knowledge as it offers better alternatives, which enable us to manage patients better
>
> **Health worker 4, Bushenyi**

The value of information has the following subcategories, as illustrated in Fig. 3.4: clinical work, administration, information dissemination, decision making and detection. The first two are discussed in this subsection, whereas the rest are discussed in the next subsection (Actions). Another subcategory that is related to the value of information is constraints to information use, which is discussed in Chapter Two (under Constraints). It was noted that in some situations, the constraints blocked information use, which hindered the attribution of value (constraints marked with a dotted line in Fig. 3.4).

Clinical work

Health workers reported that information was very valuable to their clinical work, which involved diagnosis, treatment and other formal and informal delivery of care, clinical tasks such as counseling, or the provision of advice to patients. The tasks are summarized as management of patients or clients. Some examples of the value of information:

> The short course I attended about 'Integrated Management of Childhood Illnesses' (IMCI) updated me a lot. I also got the latest edition of IMCI manual. Before then, some childhood diseases were very difficult to handle ... actually we didn't know how to manage them; so, we used to refer such cases, but there are transport challenges between here and the district hospital, and children used to die in the process! But now, I know what to do... and I have also held several sessions with my staff here at the health centre, going through the ICML manual step by step, and using some cases on the ward to apply what we discuss. We also sensitise mothers about these illnesses and urge them to rush children to health units. All in all, this health centre is now in a better position to effectively handle childhood illnesses
>
> **Health worker 5, Iganga**

Furthermore, the value of information as manifested in the critical incidents was noted, although the incidents also highlighted the various needs for information in emergencies, as presented in Chapter Two (Section 2.4.2, Information Needs). Some examples of the value and impact of information in critical incidents are given here:

> A 12 year old child was brought to the health centre bleeding from the nose profusely. After positioning him properly, I needed to find information about the reaction to one of the medicines I wanted to prescribe because I remembered from my training that this particular medicine, if not administered properly, could cause serious clotting which may result into death. So, I rushed and referred to a manual (Standard treatment guide) which fortunately provided the information about the correct dose of the medicine; I then gave it to the child and he got well
>
> **Health worker 8, Bushenyi**

> Recently I examined a patient and thought he had an intestinal obstruction; on opening, I found a ruptured gall bladder yet I had never done cholecystectomy before. I called in a colleague; before he came, I decided to read the CME materials which provided the details of what to do. When my colleague came, he brought a copy of Primary Surgery textbook by King which also gives step by step account of what to do in order to carry out that surgical operation. These assisted us to successfully do the operation. When I was relaxed after the operation, I did some online literature search and accessed even better information that I shared with colleagues, saved a soft copy and printed off and filed for future reference
>
> **Health worker 1, Lira**

Among other things, these incidents demonstrate the various types of information that are focused on in this book. For example, Health worker 8, Bushenyi got information from a manual/handbook, a physical entity, to check for the dose of medicine that he planned to prescribe, whereas other health workers such as Health worker 2, Bushenyi sought and received advice from a doctor that she then used to manage a patient with unusual herpes zoster. The two health workers used single sources of information. However, Health worker 1, Lira sought and received information from a combination of sources: continuing medical education (CME) materials and a book, as well as advice from a colleague that enabled him to perform a surgical operation, and later he accessed more evidence-based information from online databases, which he saved for future use. Information was from single or multiple sources. It was used in the management of patients and included diagnosis and treatment. Furthermore, in all these cases, health workers updated their knowledge and learned something. The critical incidents also demonstrated the centrality of the value of information as the core category because it closely relates to other categories, namely, information needs and sources.

Administration

All the heads of health units who were interviewed indicated that they produced periodic reports and work plans based on community needs assessments, health unit records and previous reports and work plans, which were in turn used for planning and provision of improved health services. Information was also reported to have been valuable in the planning and administration of health units and of health programs, as the comments show:

> The sub county health committee members also do health education in rural areas... We exchange reports which help us to plan for future health education sessions. Furthermore, the various contacts facilitate the identification and sharing of information on common areas of interest, give us a chance to compare implementation strategies and enable us to operate within the laws governing health and local councils, and hence the smooth running of programmes and projects in the community
>
> **Health worker 1, Masaka**

> Due to decentralisation, we have to liaise with district political leaders... these contacts are useful for policy, financial matters and other decisions that affect the day-to-day running of the hospital
>
> **Health worker 1, Lira**

3.3.2 ACTIONS

As it was with the women interviewees, the various actions presented in this section would not have taken place if information had not been used by the health workers. The actions directly resulted from the use and attribution of value to information by health workers. However, actions were more explicit in the women's data, and rather implicit in the case of health workers. The differences seem to stem from the fact that health workers are professionals but the women interviewees were not. In the case of health workers, there was a general professional value accorded to information; hence, there was a difference between the personal and professional actions of health workers. The personal actions (eg, behavioral change), however, showed a pattern similar to those of the women. The differences between health professionals and nonprofessionals were also manifested in the information behavior (active vs passive) such that health workers' information needs resulted in active information seeking, and what the information was used for (eg, writing a seminar paper) led to the action of dissemination of information that was facilitated by the value of information (writing of the paper made it easier for the information in the paper to be disseminated). As Fig. 3.4 shows, the subcategories under actions are information dissemination, making decisions and detection.

Information dissemination

Information was used in various health information dissemination sessions, both formal and informal.

All the health workers interviewed reported that they had various forms of informal health information dissemination sessions daily. The sessions included answering patients'/clients' questions, counseling patients/clients, providing advice to those who consulted them in or outside the clinics/health units and chatting with people during social gatherings.

In addition to informal sessions, health workers indicated that they used the information in various formal health education sessions for the general public and for specified groups such as adolescents, teachers and LC officials. Information was also used in the production of various documents and in training lower-level health workers. Information dissemination was therefore subdivided into preventive work, academic work and professional support.

Preventive work

It was noted that preventive work scored high among the uses of information by health workers. It included health promotion and education, such as sensitization/awareness raising about disease prevention and control, mobilization of the public to prevent or control other health problems by using health facilities and services (eg, FP) and the various information dissemination sessions in the communities. However, information dissemination in health units, as part of clinical tasks, was considered as part of clinical work. Preventive care was subdivided into sensitization/awareness raising, marketing of health services and facilities and valuable channel/media.

Sensitization and marketing of services/facilities were very closely related. However, sensitization was broader because it covered various health issues, whereas marketing focused on immunization, FP, antenatal and postnatal care, as well as the use of health facilities. Sensitization and marketing were therefore separated in the analytical interpretation of data. Some examples of sensitization:

> The statistics updated my knowledge and are a point of reference during sensitisation sessions …they assist me in answering questions and implementing activities in the communities… I compare our district (Bushenyi) statistics with other districts in Uganda, and use this information to sensitise communities about the need to improve e.g. nutrition of children where our district lags behind many districts in the country and yet we have plenty of milk, millet, etc… We emphasise preventive measures because curative is very expensive
>
> **Health worker 3, Bushenyi**

> MADDO (Masaka Diocesan Development Office) provides us with books, newsletters, pamphlets, and video tapes/DVDs about abortion and related topics that I use in health education. Two of the pamphlets were translated in Luganda, which makes them easy to use in this area…actually I use them a lot to counsel people; for example, in the last 2 weeks, I had two cases: a student of S6 (high school) and a married woman who wanted to abort and came to me for advice. I talked to them individually, showed them a video about the dangers of abortion and gave them

> the pamphlets to read; they came back individually and informed me that they had decided not to, and later brought the babies
>
> **Health worker 5, Masaka**

Health workers also used the information in marketing health services and facilities:

> We hold community capacity building sessions for LC3 committee members, so that they can, in turn, sensitise their immediate communities about the benefits of using health facilities and participation in health and development activities... I have to read widely about a number of issues and try to be current in my work, which gives me confidence; otherwise some difficult questions may come up and I get stranded! So, the information and knowledge I get from the different sources has been very valuable
>
> **Health worker 1, Masaka**

Various channels, media or tools were identified. They were found valuable in the preventive activities of health workers. Unlike value of information source (under clinical work/management), which referred to sources of information that health workers found valuable because they provided vital information for their clinical activities, in this section the sources were tools that assisted health workers in their activities concerning the prevention and control of diseases/health problems. They included radio, mobile phones, printed sources, drama/concerts and school children.

Academic work

Information use facilitated the production of documents and the training of other health workers. Interviewees reported that information was valuable in generating other types of information, for example, reports, seminar or conference papers and articles, proposals, training manuals, audiovisual materials, posters, charts and other print and electronic materials.

All the health workers interviewed reported that they produced periodic reports on a monthly, quarterly and/or annual basis. In addition, papers for presentation at seminars or conferences were prepared by some health workers using the information they had accessed from various sources. For example, doctors from Bushenyi and Lira reported that they had presented two papers each in the past twelve months before the interview. Some reported having written articles for publication. Other health workers used the information they accessed to prepare training manuals, whereas information had enabled some doctors to write successful project proposals for funding.

Posters, leaflets and/or charts were also produced by health workers for lower-level health units and the public. Most posters, leaflets and charts were translated in the vernacular to overcome the language barrier. They were used, among others, in seminars and health education sessions held in the vernacular.

With regard to training, interviewees reported that they used the various information accessed to train lower-level health workers such as enrolled nurses and midwives, nursing assistants, as well as the untrained community workers such as the Village Health Team members, Traditional Birth Attendants and other traditional practitioners.

Professional support

Unlike training that is formal, health workers reported that they used the health information and the knowledge and skills gained over time to train their juniors informally on the job and to answer questions or to guide them professionally in their work/activities. Furthermore, all the heads of health units reported that they shared the information during the periodic staff meetings.

In addition, all the medical doctors interviewed reported that they provided monthly support supervision to lower health units (health centers, Dispensaries/Maternity Units, etc.). This generally agrees with the findings of Wood et al. (1995: 300), which showed that "All GPs are producers as well as users of information... All the GPs produced at least some of the following information: patient health care information, updates of the patient records, patient referral letters/medical reports for other agencies, information for trainees, information for other teaching, work for continuing professional education, information for practice meetings/talks, internal practice reports, information for committees/groups and professional advice."

Making decisions

This included decisions concerning personal health, practice and related issues. However, decisions made in relation to clinical tasks remained within the clinical work subcategory. In both the clinical work and in this subcategory, information enabled health workers to make informed decisions.

Personal health included behavioral change and other general issues as illustrated in the comments below:

> Charity begins at home ... the information I receive or the knowledge I have about different health issues, first and foremost, is useful to my life and that of my family ... family health decisions are based on such information
>
> **Health worker 2, Masaka**

> I use the health knowledge I get personally to improve my health. For example, nutrition information is very valuable to me personally - I have had to change my diet following the guidelines I get from nutrition databases, and it has helped me a lot. I also use my personal experience to advise patients - especially those with hypertension and diabetes... when you talk about something you have done or experienced personally, it sounds more real, I am told... Our training didn't go into much detail about nutrition; so, the information I get now is new and very useful
>
> **Health worker 1, Iganga**

Some health workers, particularly those operating private clinics, reported that information was very valuable in making decisions concerning their professional conduct and practice and the sustainability of clinics, which were their additional source of income as the comments show:

> I heard from colleagues ... but didn't take them seriously until I read the New Vision (newspapers) and saw that clinics including some Marie Stopes units had been de-registered and closed because of carrying out abortions which is illegal by our law... It is tempting to do these things you know, because it is medically safe, and people come with such a need that may be difficult to resist... Anyway, I am now very cautious because I don't want to be in trouble ... my clinic was not even registered, so I went to Kampala for registration
>
> **Health worker 7, Iganga**

This finding differs from Witte et al.'s (1993) study, which indicated that threatening information was more effectively presented through interpersonal channels, such as conversation, than through the media, and that threatening messages given over mass media channels may simply be ignored by the audience. On the contrary, Health worker 7, Iganga's comment shows that he did not take his colleagues seriously when they discussed the harsh measures taken against illegal abortion until he read the newspapers. This may be due to a number of factors; for example, newspaper reports from or about Government ministries tend to be considered more authentic sources than colleagues. Secondly, the assumed rivalries between professionals could have made the health worker doubt the seriousness of the messages his colleagues were trying to put across because he could probably have thought that they had ulterior motives. Credibility and authenticity of information source, therefore, give added value to the use of information and the decisions that follow.

Other decisions made because of the value of information included posting newspaper clippings on health and improving the IT equipment in the health unit after an information literacy workshop by Makerere University health professionals and librarians as the comment shows:

> Before the outreach workshop was held in this hospital, only the office of the Medical superinten-dent had internet, but after the training, the hospital bought four modems to facilitate health pro-fessionals to access the internet and the online resources. The workshop was, therefore, very useful as it showed how to access current literature by searching the internet, and it made the hos-pital administration realise the need to increase the IT facilities for us.
>
> **Doctor, Iganga/Mayuge**

Detection

This involves detection of diseases and other health problems in an area. Once detected, preventive work such as health education would follow to try to prevent or control the disease or health prob-lem. Some comments:

> The information we receive enables us to monitor health issues in the district... for example, I usually contact veterinary doctors, entomologists, agriculturists, etc., for surveillance purposes... the data I get from them helps us, as a district, to predict the epidemiological trends of say, malaria and to plan accordingly
>
> **Health worker 5, Bushenyi**

> There has been a tendency to focus on rural health as if epidemics are limited to rural areas ... but the cholera epidemic that started in urban areas recently taught us a lesson... indeed epi-demics can happen anywhere anytime, we have to be more vigilant in predicting these problems; so, all information relevant to epidemic is very useful and we take it seriously
>
> **Health worker 1, Masaka**

3.4 THE EFFECT OF INFORMATION ON HEALTH CARE

The section has two subsections, namely, the background and examples of evidence.

3.4.1 BACKGROUND

There was relatively limited research evidence of the impact of information and library services in pri-mary health care in comparison with hospital settings, and the research available was generally reliant on small samples. There was a general lack of impact studies conducted with nonclinical staff (Bryant and Gray, 2006). Furthermore, Ndira et al. (2014) observed that their study was "not specifically designed to rigorously assess the impact of information on morbidity and mortality in a research context..." (p. 5.).

The assessment of the effect of information on development in general and on health outcomes in particular was, therefore, a topical issue. The major challenge was the various contributing factors that make it difficult to attribute the outcome to information alone. Among the reported improved health outcomes in Uganda, for example, was that HIV/AIDS patients were living longer in 2014 than they did twenty years ago. However, some interviewees indicated that although the increased access and use of information by both health workers and patients played a key role, the improved health outcome could not be attributed to information alone as other factors, namely, the availability of Ant-Retroviral Therapy (ART) and the new combination of drugs played an important contributing role.

Similarly, when cases of intestinal worms decreased by more than 80% in two years (2012–2014) in Iganga hospital, Eastern Uganda, the information, education and communication (IEC) activities about safe water and general hygiene as well as the Village Health Teams who distributed deworming medicines three times per year played a complementary role.

Some authors, in the field of monitoring and evaluation, recommended that the focus of measurement should be on the true, long-term and substantive impact (Lipton, 2012), which would show whether information, for example, makes a difference. Other recommendations focused on the outcomes to show, for example, the effect of an information service, activity, process or intervention on the lives and livelihoods of the community and its members. Only in 2014 did the International Organization for Standardization (ISO, 2014) Working group recommend a new International Standard, method and procedure (ISO 16439) for assessing the impact of libraries and their value to society (Laitinen, 2014).

To address this challenge, the Elsevier Foundation (2014) initiated some health impact evaluation grants for its previous project fund recipients to map health information to improved health care. The strategic objective was to create a body of coherent evidence about the long-term effects of access to scientific information and research results. The author had successfully implemented a three-year Elsevier Foundation–supported project titled "Enhancing access to current literature by health workers in rural Uganda and community health problem solving" that was concluded in March 2014. The project was part of the Innovative Libraries in Developing Countries (ILDC) program of the Elsevier Foundation (Musoke, 2014).

Furthermore, in 2015, the International Federation of Library Associations (IFLA) documented some examples of the impact of information by collecting the contributions of libraries and librarians to the sustainable development goals. Some of the examples highlighted the impact of information on health care.

3.4.2 EVIDENCE OF THE EFFECT OF ACCESS AND USE OF INFORMATION ON HEALTH CARE

Sections 3.2 and 3.3 have presented various examples of the effect of information on health care as interpreted from the raw data of the research by the author. This section summarizes the evidence of the effect of information on health care and the proposals made by the interviewees on how the effect could be measured.

Examples were narrated by the health workers giving evidence that access to and use of current literature had greatly improved the management of patients and health care in general. Some of the examples were similar to those presented in Section 3.3 (Clinical Work). The women also presented examples of evidence.

Examples of evidence by health workers

Health workers generally indicated that access to and use of current research-based evidence enabled them, among other things, to change the routines of their professional work, which improved the management of patients and the overall health care because it was more cost-effective and reduced the burden to the patients. An example:

> we used to give antibiotics for post-operative care for a minimum of five days, but different studies, some two years ago, indicated that a single prophylactic dose before operation can be, and is indeed, as effective in controlling infection; so, we changed ... and this has reduced the wastage of medicine and the burden to the patient
>
> **Doctor, Masaka**

Information was also a key factor in controlling several epidemics in Uganda and the health workers indicated that:

> When health professionals keep updating their knowledge, they are able to control the spread of epidemics e.g. the recent Ebola in Africa... Health professionals in Uganda used various information to sensitise the population to remain healthy
>
> **Clinical officer, Masaka**

Some health workers proposed a trend analysis to provide some good evidence, which also demonstrated the changes in information use behavior, as the comments show:

> I was not able to supervise my juniors qualitatively some two decades ago, but with the improved information search skills I have now, I can retrieve as much information as I need, so I am able to do so. I also attend to about 20 patients on the ward daily in addition to doing administrative work...Furthermore, the information we provide to the community and the massive immunisation have reduced measles to almost eradication levels in Sub-Saharan Africa, and Uganda is doing relatively well e.g. infant mortality rate of 54/1000 live births and the under 5 years, 90/1000 live births... Most of the MDGs have had a positive effect on child health and information access and use has been a key... In my view, a number is not a hard outcome but a soft one.
>
> **Doctor, Bushenyi**

> We applied the information we accessed and now neonatal tetanus, which was common some fifteen years ago, is almost eliminated... similarly, our hospital used to have a Measles ward, but as measles are almost eradicated, the ward is used for other cases... in fact some of the young staff in the hospital have not attended to any measles case
>
> **Doctor, Masaka**

> I read an article about better nutrition for HIV/AIDS patients and used it to propose an improvement in managing people living with HIV/AIDS who were previously only given ARVs... the addition of a nutrition component greatly improved the lives of the patients
>
> **Clinical officer, Masaka**

It was also noted that the more people accessed health information, the more they became cautious about their health, as evidenced by the number of people who went to clinics and laboratories for wellness tests when they were not sick. The following comment illustrates that point further:

> As a doctor, I have experienced a rise in the number of people who tell me that they read about something or saw it on TV and ask me where they should do the wellness tests and whether I could do the tests myself, something that was not so in the past... people waited to fall sick, then go for laboratory tests... Definitely there is a growing interest in checking for wellness.
>
> **Doctor, Masaka**

Another example was the use of a newspaper article that raised the awareness of the factory employees to demand compensation:

> there was a newspaper article on occupational health hazards for people working in battery and paint factories... the employees, after reading the article, demanded for laboratory tests and I got to know about the genesis of the problem when I was requested to carry out the tests. Consequently, the employees who had developed health problems were compensated... a perfect example of information safeguarding the health of the people
>
> **Doctor, Iganga**

The health professionals also emphasized that the various interventions in health in Uganda were based on research evidence, recorded best practices, benchmarking what worked elsewhere, and other practices that they read about and were modified to local conditions and implemented. Such interventions greatly contributed to improving the overall health in the country.

The use of mobile devices to consult senior colleagues, to check the Internet or to call the area hospital for the ambulance was also reported to have saved lives in many situations, for example:

> lower level health units, which did not have phones in 2000, now use mobile phones to call the hospital to send the ambulance to pick patients, which has improved the health outcomes
>
> **Nurse, Iganga**

Examples of evidence by the women

The examples of evidence identified by the women are summarized in this subsection. Women indicated that the increased access and use of information had greatly reduced the HIV/AIDS disease stigma in most parts of Uganda:

> Stigma has reduced by about 80% in this area... more people now talk openly about their HIV status ... they discuss and share experiences which help others...counselling has provided the information needed to keep people going and avoid the many suicidal cases of the past ... people ask openly when the next supplies of ARVs will be made, when the next blood test will be done, etc which was not the case twenty years ago when people used to hide and manage the AIDS problem in secrecy, but without success!
>
> **Woman 7, Iganga**

Information about HIV testing during pregnancy and mother-to-child treatment also led to a reduction in babies born with HIV, which was confirmed by health workers and hospital records. This had increased the number of pregnant women attending antenatal clinics and delivering in health units. It was further reported that women who delivered in health units also had their babies immunized, which improved health. An immunization card was one of the requirements before admission to most Nursery and Primary Schools. An increase in the immunization of babies was reported to have led to a reduction in the disease burden of immunizable diseases such as measles.

Furthermore, information, through awareness raising, led to the eradication of small pox. The interviewees from Iganga district, where the small pox problem had been rampant, reported that since the eradication of small pox, the community values information because it was the main strategy used. They also referred to the Ebola problem and indicated that the information provided by the Ministry of Health (MoH) enabled Uganda to prevent the epidemic in 2014.

Proposed ways of measuring the effect of information on health care

The following proposals were made for measuring the effect of access and use of information by health professionals on health care:

i. The number of postoperative infection cases or rates (the evidence-based literature used by health professionals led to the control of the infections before the surgical operation in a specified hospital);

ii. The number of drugs consumed/administered to patients and the cost-savings (as a result of changing the management of patients using the evidence from research information that provided better treatment options);

iii. The growing trend in the number of people seeking wellness tests such as blood sugar levels and hypertension (resulting from the information provided by health professionals or accessed by the public from various sources);

iv. The reduction in the number of health professionals who manage elite patients suggesting treatment options that they (health professionals) were not aware of (given that the health professionals are more informed and knowledgeable);

v. The improved administration, planning and budgeting for health services at different levels (using information and improved experience to overcome wastage and reduce the burden on patients);

vi. The performance of medical students in a specified period of time and the quality of theses/dissertations (resulting from the guidance by better-informed lecturers/supervisors);

vii. The quality of articles/papers written by health professionals and how they rate globally among the professional peers (given the increased access to current literature);

viii. Based on health indicators, perform periodic audits to get the patterns and trends of disease occurrences (because both the health professionals and the public have improved access to information).

With regard to how the changes in information use could be measured, health professionals proposed various measures in a given period of time, for example: the number of patients treated using evidence-based practice rather than what one learned in the medical school; papers/articles on a specified topic, downloaded in a specified period and read; the frequency of literature searches/usage statistics of online databases, including HINARI; papers published by a specified number of health professionals; CME/continuing professional development (CPD) presentations then and now (used to be based on older versions of printed sources but now are on current evidence-based research during the year); health professionals with smartphones and/or other mobile devices using the Internet to access scientific health literature and health professionals with modems and/or routers for Internet access.

Furthermore, the speed at which one retrieves current scientific information, the ease with which one communicates by sending an e-mail to request for literature, the length of time HIV/AIDS patients live after infection, clinical outcomes now and in the past, and qualitative assessment of health professionals by hospitals would shed more light on the value of information and the changing information seeking and use behavior of health workers.

The women also proposed various ways of measuring the effect of information access and use on health care by the communities they lead or the public in general. The proposals include:

i. The reduction in infant mortality and morbidity in Uganda (due to the increased information access and use through community mobilization and sensitization as well as health education as the major factors). For example, a reduction in infant mortality rate, reduction in the number of children born with HIV, as well as a reduction in children dying from immunizable diseases in a specified period of time. The author confirmed the reduction in infant mortality and morbidity between 2000 and 2014 from the Uganda Demographic and Health survey report.

Using children's health as a yardstick for mothers who attend health education/promotion sessions, the number of times their children suffer from preventable diseases such as malaria, intestinal worms, diarrhea and immunizable diseases, which were focused on during the training sessions, should show the effect of information on health care, as the comparison before and after the information sessions would reveal.

ii. The increase in the number of women attending and delivering in health units (after mobilization and sensitization about the importance of delivering in health units). The measure should show the number of mothers attending antenatal clinics in a given district and period of time and delivering in health units.

iii. The reduction in the number of cases with intestinal worms in Iganga district (as a result of community mobilization and sensitization about drinking boiled water, general hygiene and using deworming medication periodically). This was confirmed by the Iganga hospital records.

iv. Reduction in cases of suicide among AIDS patients (due to the increased provision of information through counseling both the patients and the carers).

v. Increase in the number of households with improved hygiene and, consequently, reduction in the number of times one sought treatment (as a result of applying the information accessed by taking preventive measures that improved health and led to, among other things, savings on household health expenditure). So, the measure would include: how often a person suffered from a particular preventable disease; the amount of medicines consumed over a period of time in a specified area and number of people seeking health care in health units in a specified period of time (after being mobilized to do so).

3.5 DISCUSSION OF THE VALUE AND IMPACT OF INFORMATION

The meaning that information has to people after being accessed, used and interpreted and its significance and role as perceived, experienced and reported by the interviewees were conceptualized as the value people attributed to information. The value of information in this book was quite subjective, although it was shared with others, as indicated in Sections 3.2.2 and 3.3.2 (Actions). Hence, this book's approach to the value of information is what Saracevic and Kantor (1997) referred to as "perceived value approach," which is a "subjective valuation by users of information, of the value or benefits of the given information. This assumes that users can recognise the value of information (or the benefits gained/lost). If scales are used, it assumes that they can place the value in some ranking" (p. 532).

That approach differs from the normative and realistic value approaches also discussed by Saracevic and Kantor (1997). The normative value approach, for example, takes the narrowest view of information because it excludes the broader attributes and aspects. As indicated in chapters "Health Information in Uganda" (Section 1.2) and "Access and Use of Information by Women and Health Workers" (Section 2.1), this book has a broader perspective of information. Although the realistic value approach is less restrictive, both the realistic and normative approaches assume information as an exclusive, identifiable variable and share the difficulty of resolving it from other variables that also affect the process of decision making. These authors indicated that the normative approach had not yet been successfully applied in theory or practice related to the value of library and information services. The realistic approach, however, was reported to have been applied generally in the economics of information, and several studies on economics and/or the value of library and information services were reviewed by Feeney and Grieves (1994). Others focused on measurement of information (Tague-Sutcliffe, 1995; Menou, 1998; Menou and Taylor, 2006).

There was also a value of information (VoI) analysis method used in various health technology assessment projects, which was cost-oriented (Claxton et al., 2004; Eckermann et al., 2010). Furthermore, the VoI method was used in the field of space. For example, Laxminarayan and Macauley (2012) indicated that the VoI method was critical in informing decisions on investment in satellites that collect data about air quality, fresh water supplies, climate and other natural and environmental resources affecting global health and the quality of life.

The perceived value approach, as this book has shown, is not restrictive and it brings in what Saracevic and Kantor (1997) referred to as a "collection of dissimilar attributes." The major advantage is that this approach is based on the judgments of the users who, as recipients of information, are able to decide whether the information they accessed and used was valuable or not. Information use and the value of information, therefore, are not synonymous because some of the information that interviewees used was not considered valuable due to a number of reasons, such as being irrelevant or inapplicable, as indicated in Chapter Two (Constraints section).

It was not until the late 1990s that studies in evidence-based medicine highlighted the unquestionable value of information. Before then, the value of information was considered from the economic point of view. Davies (1991), for example, indicated that "most so-called information is, in fact, merely an expense until it is used...At all stages the most important question is, how can the information professional add value to their corporation and at what cost?" (p. 48).

The savings and other tangible benefits organizations accrued after using information dominated the literature concerning the value of information for some time. Such economic aspects were used to justify the employment of information professionals and the establishment of information centers in organizations/institutions other than academic ones. However, even in the 21st century and in academic institutions, librarians were still struggling to justify the value of the library from the economic perspective using various measurements such as the return on investment (RoI) method. For example, Berg (2012) indicated how RoI failed or killed the academic library and discouraged reducing the value of libraries to what he referred to as a "numbers game," and yet there is a non-measurable impact of academic libraries. Furthermore, Abram (2010) indicated that the value of academic libraries was often tied to the impact of universities and other tertiary institutions on the economics of a community. Namhila (2015) carried out a study that focused on archival materials and reported that documents retained value as they transited from active records to archives and finally to rare materials in registries, libraries and archival centers. That study was important but different from the focus of this book.

Furthermore, the 2015 World Library and Information Congress' joint session by the International Federation of Library Associations (IFLA) Academic and Research Libraries' Standing Committee, Management and Marketing Standing Committee and the E-Metrics Interest Group was titled "What is value?" Papers were presented by academic, research and public librarians. This further confirms that value of information is a topical issue among information professionals as well as other stakeholders. The findings presented in this book, which have elaborated on the value of information in the lay women's lives and health workers' professional and personal lives, are therefore novel.

For health workers, the value was reported mainly from the professional rather than personal point of view. Some examples were given when health workers accessed information that enabled them to make decisions concerning personal health, such as condom use and diet. The women's data, however, revealed that the value of health information was mainly for the family, then personal, and community, in that order. Hence, the value of information in this book was mainly at two levels: the social level, because it served communities, and an individual level. However, the two levels were interdependent. Health workers shared their individual valuable experiences with patients on issues such as diet or other professional matters with colleagues. Women leaders did the same.

Keeping in mind that an interview is a social situation that may influence the interviewee's expressions and the reporting of events or what happened, the researcher/author sometimes doubted what she got from the field. This was because it was one thing for women leaders to report, for example, that they passed on information to their fellow women and the community in general, or for senior health workers to report their support supervision, but it was yet another for short-term research to be able to confirm that information was actually disseminated by this particular interviewee, or that health workers benefited from support supervision provided by a particular doctor. However, in several instances, the author was able to confirm what had been reported. In Bushenyi district, for example, the first woman who was interviewed was the LC1 secretary for women; she narrated her various dissemination activities, and then other interviewees (namely, Woman 2, Bushenyi) confirmed the dissemination meetings held by the secretary for women as a valuable source of health information. Similarly, Health worker 4, Iganga confirmed the valuable weekly support supervision by Health worker 8, Iganga (as indicated in Chapter Two, Moderators to Information Access, Section 2.4.4).

Apart from showing the information flow between interviewees, this highlights the linkages between the value of information and moderators. It also demonstrates that some moderators to information access also moderated information use. As women leaders or senior health workers provided information to those who had been constrained to access it, they interpreted or repackaged the information and passed it on in a way that would render it usable.

Hence, the findings have demonstrated information as a commodity because, among other things, it gained value as it passed along the production chain. This was contrary to what was indicated in Chapter One (Section 1.2) that this book does not subscribe to the notion of information as a commodity. In addition, the social aspects of society and human behavior facilitated information access and use more than the economic aspects, which were reported to have constrained many interviewees from accessing information. This finding is similar to debates concerning modern social development, that is, whether the modern world has been shaped by economic factors or by other influences, such as social, political or cultural (Giddens, 1997, 2013).

The value of information in the prevention of diseases and promotion of health has been clearly demonstrated in this chapter. This book argues that the value of information, rather than needs or constraints, was the driving force behind the information processes reported. The various actions (Sections 3.2.2 and 3.3.2) that resulted from the value of information were reported to have promoted health in many ways. This agrees with the World Health Organization, which reported that "both the public health and the personal care interventions have contributed to reversing the urban—rural differences in health status; better health among urban populations is due more to the application of improved knowledge than higher incomes in cities" (WHO, 2000: p. 10).

It follows, therefore, that although rural areas had low incomes, they could enjoy better health if they accessed information to enhance their knowledge. Hence, factors that negate information access and use in rural areas need to be addressed so that rural communities may reap the benefits of improved health knowledge.

As indicated in Chapter Four, a core category should be central, that is, it should relate meaningfully and easily to as many other categories and their properties as possible. An example of inter-relationships arising from the constant comparisons between and within categories to show the centrality of the core category—value of information—and how it relates to, for example,

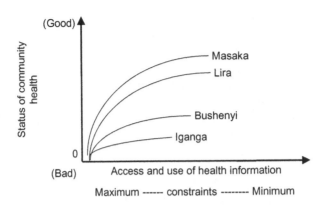

FIGURE 3.5

The relationship between the extent of information accessibility and use and the status of health.

information needs and sources in the case of health workers revealed that health workers used a combination of sources (eg, seniors or colleagues, seminar notes, textbooks/handbooks and/or online databases) for a single information need. Some of the information they received and used filled gaps and solved their clinical, planning, academic, preventive and personal health problems or information needs. The value of information drove the activities as health workers sought information because of its known or anticipated value. So, although an information need could trigger an information process, it was driven and sustained by the value of information.

There was some difference in information access and use in the study districts. Ideally, the more information was accessed and used, the higher the resulting status of health, as the WHO (2000) report cited above showed that "better health... is due to the application of improved knowledge." The status of community health was the impact, wheras the value of information and moderators were the major actors that intercepted, stopped or solved glaring bottlenecks or constraints to achieve the improved status of community health.

The women's findings in Chapter Two (Section 2.3.2, Information that was Accessed) showed that Masaka had the highest awareness about the identified diseases/health topics, followed by Lira, Bushenyi and Iganga. Information use had a similar pattern. Other things being equal, this would imply that Masaka and Lira experienced a relatively better status of community health than Bushenyi and Iganga, as illustrated in Fig. 3.5. The figure also shows that the more the constraints are reduced to a minimum, the more would information be accessed and used.

Access and use of information were measured against the resolution of constraints; the values of information and moderators were the major contributors to that resolution. The farther away one moved from the point of origin, the less were the constraints brought about by the intervention of the value of information and moderators, and the greater was the access and use of health information. This shows not only the centrality of the value of information but also how, as a core category, it fulfills the function of solving or resolving the problem of information access and use in rural Uganda.

3.6 CONCLUSION

This chapter has demonstrated, among other things, that the use of research-based information as evidence had greatly improved the knowledge base of health professionals, which led to better clinical outcomes than ever before. There were clear cases in which scientific information brought about improved health outcomes, for example, when treatment was changed using scientific evidence that reduced wastage of medicines and the burden to the patients. This can easily be measured by the number of postoperative infection cases or rates (because the infections were controlled before the operation). Furthermore, various cases of improved health outcomes were presented by the women as a result of community mobilization, sensitization and health education.

At a country level, Uganda reported repeatedly that information was the major line of defense in the battle against HIV/AIDS, and several authors (Green et al., 2006; Hogle, 2012; Omaswa and Crisp, 2014) emphasized that Uganda had used IEC to successfully reverse the prevalence of AIDS. The same strategy was used to control Ebola in 2012 and to prevent it from affecting Uganda in 2014 after reporting it in the neighboring countries such as Congo.

Generally, the book focuses on people's everyday life health-related information situations in a holistic way and assesses the whole information situation. The use, interpretation and meaning of information reported by interviewees gave prominence to the value and effect of information as the driving force in their information activities.

The assessment of the effect of information on health care has been an issue of concern to many development agencies and planners. The chapter has presented the many positive effects of access and use of information on health care in a low-income African country. There was hardly any research-based and comprehensive literature on this topic in Sub-Saharan Africa. Even in other parts, most literature about the value of information focused on the immediate economic benefits rather than the long-term effect on the lives and livelihoods of communities such as those presented in the chapter. The findings are, therefore, novel because they would enhance our understanding of this important topic.

The next chapter introduces and presents a new information model that advances our understanding further. The model is one of the major contributions to knowledge by this book.

REFERENCES

Abram, S., 2010. Value of academic and college libraries. <http://stephenslighthouse.com/2010/04/07/value-of-academic-and-college-libraries/>. (accessed september, 2015)

Berg, J., 2012. How ROI killed the academic library. <https://chrisbourg.wordpress.com/2012/12/18/how-roi-killed-the-academic-library/>.

Braman, S., 1989. Defining information: an approach for policy makers. Telecomm. Policy 13 (3), 233–242.

Brettle, A., 2015. Measuring the impact of health libraries in practice. EAHIL conference. <https://eahil2015.wordpress.com/workshop-presentations> (accessed 30.07.15).

Bryant, S.L., Gray, A., 2006. Demonstrating the positive impact of information support on patient care in primary care: a rapid literature review. Health Inf. Libr. J. 23 (2), 118–125.

Claxton, K., et al., 2004. A pilot study on the use of decision theory and value of information analysis as part of the NHS Health Technology Assessment programme. Health Technol. Assess. (Rockv) 8 (31), 1–103.

Davies, E., 1991. Expanding Horizons: the Information Professional and Management. ASLIB, London.

Defleur, M.L., Dennis, E.E., 1989. Understanding Mass Communication. third ed. Houghton Mifflin, Boston.

Eckermann, S., et al., 2010. The value of value of information: best informing research design and prioritisation using current methods. Pharmacoeconomics 28 (9), 699–709.

Elsevier Foundation, 2014. Innovative libraries in developing countries (ILDC) <http://www.elsevierfoundation.org/innovative-libraries/>.

Feeney, M., Grieves, M., 1994. The Value and Impact of Information. Bowker-Saur, London.

Giddens, A., 1997. Sociological Theory. third ed. Blackwell Publishers, Oxford.

Giddens, A., 2013. The Consequences of Modernity. Polity Press., Cambridge.

Ginman, M., 2000. Health information and quality of life. In: Dowd, C., Eaglestone, B. (Eds.), Health Information Management Research. Proceedings of the Fifth International Symposium—SHIMR, June 12–13. University of Sheffield, Department of Information Studies, Centre for Health Information Management Research, Sheffield, pp. 9–19.

Glaser, B., 1978. Theoretical Sensitivity. Sociology Press., Mill Valley, CA.

Gray, P., 1991. Psychology. Worth Publishers, New York, NY.

Green, E., et al., 2006. Uganda's HIV prevention success: the role of sexual behavior change and the national response. AIDS Behav. 10 (4), 335–346.

Haythornthwaite, C., 1996. Social network analysis: an approach and technique for the study of information exchange. Libr. Inf. Sci. Res. 18, 323–342.

Hogle, J., 2012. HIV prevention programs before the introduction eHealth in Zambia and Uganda. < https://ehealthutsc.wordpress.com/about/2-hiv-prevention-programs-before-the-introduction-ehealth-in-zambia-and-uganda/ > (accessed 15.10.15).

International Federation of Library Associations (IFLA), 2015. Call for examples and how do libraries further develop. http://www.ifla.org/node/9830 http://www.ifla.org/node/7408.

International Standards Organisation, 2014. ISO 16439—Information and Documentation—Methods and Procedures for Assessing the Impact of Libraries. ISO Copyright Office, Geneva.

Laitinen, M., 2014. Can impact be standardised? ISO 16439 Standard as a new tool for evaluating the impact and value of library. < http://www.isast.org/images/FINAL_BOOK_OF_ABSTRACTS_e-book_version.pdf >.

Laxminarayan, R., Macauley, M.K. (Eds.), 2012. The Value of Information: Methodological Frontiers and New Applications in Environment and Health. Springer, Dordrecht and New York, NY.

Lipton, D., 2012. What defines a modern library [Blog, August 28]. Available: < http://irexgl.wordpress.com/2012/08/28/what-defines-a-modern-library-exciting-conversations-emerging-from-the-international-young-librarians-academy-in-ventspils-latvia/ > (accessed 15.09.15).

Menou, M.J., 1998. Does information make any difference? Br. Libr. Res. Bull. Autumn (21), 10–12.

Menou, M.J., Taylor, R.D., 2006. A grand challenge: measuring information societies. Inf. Soc. Int. J. 22 (5), 261–267.

Mundell, E.J., 2000. Uganda first in Africa to show drop in new HIV infections. In: Proceedings of the XIII International AIDS Conference, Durban, South Africa. < http://dailynews.yahoo.com/htx/nm/20000713/hl/uganda_hiv_1.html >.

Musoke, M.G.N., 2014. Enhancing access to current literature by health workers in rural Uganda and community health problem solving. < http://library.ifla.org/868/1/088-musoke-en.pdf >.

Namhila, E., 2015. The dilemma of value as a concept. < http://www.ifla.org/node/9847 >.

Ndira, S., et al., 2014. Tackling malaria village by village: a report on a concerted information intervention by medical students and the community in Mifumi, Eastern Uganda. Afr. Health Sci. 14 (4), 882–888.

Omaswa, F., Crisp, N., 2014. African Health Leaders: Making Change & Claiming the Future. Oxford University Press, Oxford.

Saracevic, T., Kantor, P., 1997. Studying the value of library and information services. J. Am. Soc. Inf. Sci. 48 (6), 527–542.

Tague-Sutcliffe, J., 1995. Measuring Information: An Information Services Perspective. Academic Press, San Diego, CA.

WHO, 2000. Health Systems: Improving Performance. The World Health Organisation (WHO), Geneva (The World Health Report 2000).

Wilson, T., 1981. On user studies and information needs. J. Doc. 37 (1), 3–15.

Witte, K., et al., 1993. Testing the health belief model in a field study to promote bicycle safety helmets. Commun. Res. 20, 564–586.

Wood, F., et al., 1995. Information in primary health care. Health Libr. Rev. 12, 295–308.

MODELING INFORMATION BEHAVIOR

4

4.1 INTRODUCTION TO THE MODEL

Within the process of data analysis, there was coding that involved identifying, selecting, cutting-and-pasting and categorizing, plus the inevitable reading and re-reading of the data and the concepts and categories that emerged from the data. During that process, a number of thoughts, ideas and questions about the categories and their dimensions and the phenomenon being studied arose. They included the relationships as well as variations between and within categories. The ideas and questions were recorded, as they arose, in the researcher's "notes" folder, indicating the category or interview question(s) they referred to, and sometimes the verbatim quotes from the field notes. Several diagrams were also sketched to represent the thoughts and ideas. These may be referred to as memos and diagrams according to various qualitative research analysis authors, for example: "Memos represent the written forms of our abstract thinking about data. Diagrams, on the other hand, are the graphic representations or visual images of the relationships between concepts… Memoing and diagramming are important elements of analysis and should never be omitted, regardless of how pressed the analyst might be for time" (Strauss and Corbin, 1990: 198). Furthermore, "the bedrock of theory generation is writing of theoretical memos. If the analyst skips this stage by going directly to sorting or writing, he is not doing Grounded theory. Memos are the theorising write - up of ideas about codes and their relationships as they strike the analysts while coding. Memo writing is a constant process that begins with the first coding of data, and continues… to the very end" (Turner and Martin, 1986: 151).

As time went on during the analysis, the notes or memos and diagrams stimulated further and deeper thinking and creativity that culminated in theory development. A model of health information access and use in rural Uganda finally emerged. It is a multivariate model with two root and three emergent categories.

4.2 THE CORE AND MAIN CATEGORIES

The model, which was inductively derived from data analysis, had emergent and root categories that formed five preliminary categories: information sources, information needs, constraints, moderators and value of information. The last three categories emerged through Grounded theory analysis, whereas information sources and needs were root categories, which the researcher started with, as outlined in Chapter Two (Section 2.2). The root categories originated from previous studies, as highlighted in Chapter One. However, what came out of the root was derived inductively from data.

Informed and Healthy. DOI: http://dx.doi.org/10.1016/B978-0-12-804290-8.00004-X

The main categories are presented before the core category. The information behavior is also highlighted. Although the categories were defined in Chapter Two, they are outlined in this section to ease reference.

Moderators: These are aids or agents that act to overcome or reduce constraints to information access and use. Without their intervention, information processes could be halted by constraints.

Constraints to information access: These occur between the recognition of a need for information (information need) and the source from where information would be sought for in the case of active seekers. For passive access, constraints occur between information sources and the person who could access the information (as shown in stage 2, Fig. 4.2). When the constraints are not overcome, information is lost or not accessed.

Constraints to information use: These intervene after information acquisition. When they are not overcome, information or knowledge is not put to use or applied.

Information sources: This is where the information was obtained from (actual) or could be obtained from (potential). Information sources exist even when there is no apparent (active) need for information. For the women, most information was accessed passively, whereas some information was obtained through active seeking. The reverse is true for the health workers. Hence, women and health workers interacted with sources passively or actively for latent or apparent needs.

Information needs: The apparent need for information makes people seek for it (information) actively from the information sources; however, for latent needs, people access information passively and then may realize that they need it. Hence, from sources, information 'goes' to latent and apparent needs; the unmet needs may lead to active information seeking from sources.

Furthermore, unmet information needs may become constraints to information use and vice versa (as indicated in Chapter Two, Section 2.3.3, Constraints to Information Use-Attitudes and Views, and Section 2.5.3, Constraints); Wilson (1981) also reported that unsatisfied information use leads to need.

Value of information and actions: This is the value attributed to information as perceived, experienced and reported by the interviewees. The value of information is the dominant category in the model. It triggers or facilitates various actions. Some actions reduce or impact constraints to information access; for example, information dissemination becomes a source of information or a moderator. The people who had been constrained to access information in one way would access it in another way (and the series of processes of information access and use go on as value-added information leads to further interactions). For example, when women received information, used it and found it valuable, they conducted various information dissemination sessions, both formal and informal, as they interacted with their networks. The value and impact of information also made health workers disseminate information to colleagues and other people in various ways. These information dissemination activities, therefore, were driven by the value of information and involved interaction with individuals, groups and communities in the case of women leaders, or fellow health workers and patients in the case of health workers. Hence, the Interaction-Value model that emerged from the findings.

Initially, five main categories made up the model, but the main theme or the substance of the data is that "*access and use of information is a series of processes that depends on the value and impact of information to overcome or reduce constraints.*" Hence, the value of information is the core category and together with the other four categories, the model was formed. This was consistent with a recommendation made by Glaser (1978) and Strauss and Corbin (1990) that a model should have one core and a few other categories that relate to it: "The generation of theory occurs around a core category... Without a core category, an effort of Grounded theory will drift in relevancy and workability... Most other categories and their properties are related to it... Upon choosing a core category, the first delimiting analytical rule of Grounded theory comes into play: only variables that are related to the core will be included in theory. Another delimiting function of the core category occurs in its necessary relation to resolving the problem... Without a focus on how the core resolves, solves or processes the problem, the analysis can drift... instead of being forced to integrate around the problem. Yet another delimiting function of the core category is its requirement that the analyst focus on one core at a time" (Glaser, 1978: 93).

Although the constraints appeared to be dominant in the data, they did not fulfill the second function of a core category, namely, solving the problem of information access and use in the study areas; instead, they negated the process and aggravated the situation. The value of information, however, triggered various actions that impacted a number of constraints, thereby enabling people to access and use information. Many constraints emerged from the data, but so did the value of information. In view of the identified constraints, it could be argued that if it were not for the value of information, there would be hardly any information accessed and used in the rural areas studied.

Finally, through data analysis and interpretation, the two root categories formed *one main category*, namely, interaction with sources passively or actively for latent and apparent needs. However, the three emergent categories formed *one main category*, namely, moderation of constraints, and *the core category*, namely, value of information. The main and core categories are elaborated further in Section 4.3.

4.3 DEVELOPMENT OF THE MODEL: FROM CONCRETE TO ABSTRACT

Although the study of women and health workers derived a model with a core and two main categories, as outlined in Section 4.2, the model for women differed in detail from that derived after the analysis of health workers' interview data. The major differences were in the details of each category that make up the model and in the information behavior. For example, the analysis of data from the women revealed that they mainly accessed information passively, except in critical situations when they actively sought for information. Although the women passively accessed information in most situations, the subsequent user behavior was active. The women's information behavior was, therefore, dynamic. This was elaborated in Chapter Two and Three when, for example, the women's information needs changed from latent to active and vice versa. The analysis of data from the health workers, however, revealed a different information behavioral mode. Health workers mainly accessed information through active seeking, but some amount of health information was also accessed passively.

The differences in detail stemmed mainly from the health workers' professional responsibilities, roles and activities, which are different from those of ordinary women leaders in the same rural areas. For women, the process of information access and use was dependant on the relationship and interaction between their social and information environment in everyday life, whereas for the health workers, professional matters added a further dimension to their information activities.

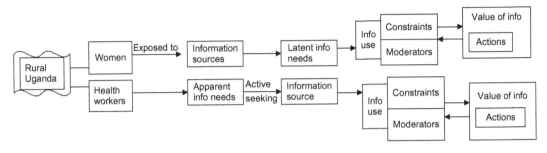

FIGURE 4.1

Stage one.

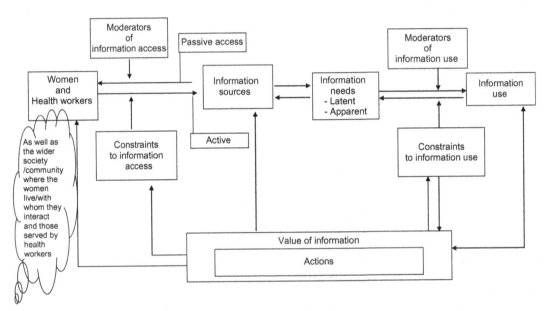

Note: Between information use and value of information, the arrow goes both ways, because information use leads to value of information and vice versa.

FIGURE 4.2

Stage two.

However, the initial models for women and health workers (Fig. 4.1) both had the same core and main categories, which shows the internal coherence of the models and the general consistency of the findings. This reinforces the fact that the research addressed the key issues concerning health information access and use, thereby providing support for confidence in the general validity of the overall model (Fig. 4.4). The development of the model moved from a concrete to an abstract situation through a number of stages, as summarized and presented in Figs. 4.1–4.4.

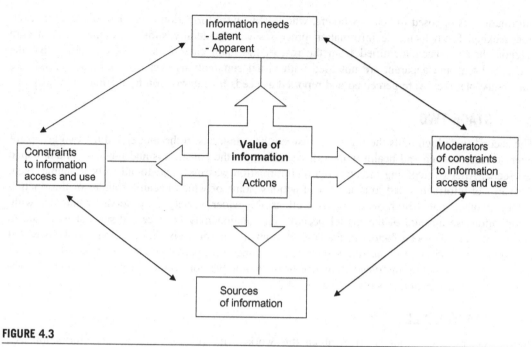

FIGURE 4.3

Stage three.

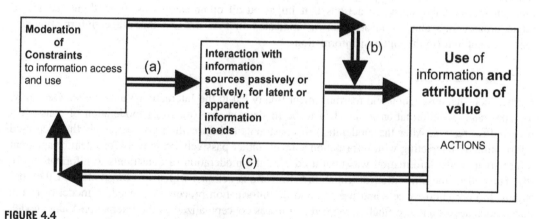

FIGURE 4.4

A diagrammatic representation of the Interaction-Value model.

4.3.1 STAGE ONE

This stage presents the main categories and the general pattern of information flow, highlighting the differences between the information behavior of women and health workers (Fig. 4.1). The findings showed that women's information needs were mainly, but not exclusively, latent until exposed to information, as opposed to health workers, whose apparent information needs led to active information seeking. From then, the information processes were generally similar. At this stage, a core category had not been identified from the rest of the main categories. (It is also noted that the concepts latent and apparent are not used with visual connotations because information needs are not observable; they were perceived and reported as needs for information by the interviewees.)

4.3.2 STAGE TWO

The second stage highlights the linkages between the categories in the model and the two groups of interviewees (women and health workers). At this stage, the constraints and moderators dominated the scene. After overcoming the constraints, information was used and found to be valuable; then, the value of information led to a number of actions, some of which became sources of information for the women and health workers, as well as the wider society they serve or interact with. Information use appears in the model because information may be accessed but not used due to constraints such as social factors in the case of women. Alternatively, information may be used but not found valuable due to constraints such as irrelevance or inapplicability. The moderators and/or the value of information intervene to reduce or overcome the constraints, thereby ensuring the continuity of the information processes (see Fig. 4.2).

4.3.3 STAGE THREE

Among other things, what is novel about this work is its qualitative focus on information use, which is a step beyond information seeking and acquisition or retrieval, where many information science studies had been stopping. The findings revealed the value of information as the driving force in the model resulting in actions that impacted all other categories. Everything seemed to rotate around the value of information as indicated in Fig. 4.3. At this stage, information use remained implicit (in the value of information).

4.3.4 STAGE FOUR

Further thoughts, abstraction and rearrangement finally led to the Interaction-Value model. Generally, the constraints to information access had to be moderated first (a). Most moderation of constraints involved interaction. After the moderation of constraints to information access, people then accessed information by interacting with information sources, either passively or actively for latent or apparent information needs. Information was then used after the moderation of constraints to information use (b). This led to the attribution of value and the various actions, which in turn moderated the constraints to information access and use (c), and the information process continued. A model of health information access and use finally emerged and it was conceptualized as the Interaction-Value model. That overall model attempted to incorporate the four stages and is presented in Fig. 4.4.

4.4 DISCUSSION OF THE MODEL

The Interaction-Value model consists of a set of related categories that, taken together, can be used to explain the phenomenon of health information access and use in rural Uganda. It is an action-oriented model that is consistent with what Strauss and Corbin (1990: 123) indicated: "Grounded theory is an action oriented model, therefore in some way the theory has to show action and change, or the reasons for little or minimal change."

It is also a grounded model because the concepts that formed it were derived from the women's and health workers' own expressions or responses, and it (the model) has been modified as data analysis and interpretation progressed. Hence, although the study, on which this book is based, was more applied than basic, concepts emerged out of the research context, with the exception of the two root categories (information sources and needs). All the other concepts were not decided at the outset of the research. This was different from what Bryman and Burgess (1994: 219) indicated: "When research has an applied emphasis, and perhaps especially when it is externally funded, the need to focus on certain concepts which are decided at the outset of the research is more pronounced... there can be discerned a greater recognition of a need to ensure that certain topics are addressed. This tendency is likely to mean that some concepts are 'given' at the outset of the research."

The model presents a process of human information behavior involving cognitive, affective and contextual factors. However, some authors had argued that speaking of a behaviorist paradigm or framework in user studies can be misleading "because there is a growing body of research which focuses on users' experiences or sense-making practices and sees these as the essential phenomena to be explained (e.g. Kuhlthau, 1993). These could be called in-between approaches; approaches which are constructionist, but not explicitly social constructionist. They differ from the tradition of behavioural science which explains information behaviour within a model in which independent variables influence dependent variables causally through particular mechanisms (if A, then B)" (Talja et al., 1999: 759).

As users interact with sources (of various types: human/oral, visual, printed, electronic, etc.) and make meaning or sense out of the interaction, what Talja referred to as "in-between approaches" could be termed "interactionist approaches." In this book, the interaction emerged from the analysis and interpretation of users' experiences as reported in the interview situation. Furthermore, although the study did not have causal approaches, the interaction that emerged from the data is clearly part of information behavior.

A number of authors discussed the need to make women more visible in existing theories and pointed out that women's experiences as individuals and as social beings, their contributions to work, culture, knowledge, their history political and other interests had been systematically under-represented by mainstream discourses in different disciplines (Narayan, 1989; Mutua, 1997). Later, Ilie et al. (2005) pointed out that there seemed to be differences between men and women in health information seeking behavior and that gender may be a factor influencing attitudes toward technology and information seeking. With regard to methodological disparities, Dervin et al. (2006) suggested that research should be four-dimensional: philosophic examination of assumptions, substantive theorizing about the real, methodological consideration of the means of the concrete, and competent and systematic execution of method.

This work attempted to bridge that gap in the information studies field. It focused on the lower levels of the primary health care (PHC) set-up, as indicated in Chapter One, in which women are care providers within the African family. Furthermore, the qualitative approach used in the research made it possible to study women's personal perspectives and experiences. Hence, the model and the rest of the findings have documented the rural women's information activities and behavior in the health sector.

The health workers also stressed the importance of and justified the need for improved provision of information to women because lack of or delayed interpretation of illness symptoms caused delays in seeking appropriate treatment, which resulted in loss of lives in some situations. This was supported by some authors who pointed out that relevant health behavior was governed as much by access to resources as by information or knowledge of appropriate prevention or treatment. The authors consider "the tendency to underestimate women's roles and skills in treatment and caring to be itself a major obstacle to improved health care... As well as preventing disease through provision of basic needs in the home, women are likely also to have more or less specialised knowledge of disease symptoms, determining when family members are ill and what kind of care they should be given. It is usually women who prescribe remedies, decide at what point in an illness to seek outside attention, and what type of practitioner to consult" (Wallman, 1996: 12).

It was equally important to study health workers at the grassroots because most previous studies had focused on district health teams or hospitals, as Chapter One indicated. The few studies that focused on rural areas (for example, Apalayine and Ehikhamenor, 1996; Patrikios, 1985; Kapiriri and Bondy, 2006; Paek et al., 2008) made recommendations that necessitated this in-depth and more interpretative work. The findings, therefore, are novel and shall enhance our understanding of this important topic.

The findings have also confirmed what had been revealed by the literature, that there were hardly any appropriate rural health information models to emulate. As a contribution to the field of information studies, an Interaction-Value model has emerged from the study of access to and use of health information in rural Uganda.

The model is driven by the value of information, unlike many previous information models (Ellis, 1993; Wilson, 1997) that give prominence to information needs as the driving force. This may be due to lack of or limited focus on information use, as Wilson (1997: 569) pointed out: "Finally, the model needs extension to include information processing and information use, which are the stages beyond information—seeking and which provide the link back to the needs."

Authors who focused on information use include Saracevic and Kantor (1997) (acquisition-cognition-application model of information use) and Dervin and Nilan (1986) (situation-gap-use model), and several other authors focused on the knowledge gap (Bonfadelli, 2002; Eveland and Scheufele, 2000; Guttman and Salmon, 2004). They all focused on information needs as the major drivers. However, the Becker Medical Library model reported by Shore (2015) focused on the assessment of the research impact on health workers. Earlier on, Johnson and Meischke (1991) used utility and actions as individual and separate categories. In this book, actions resulted from the value of information and therefore could not stand as a separate main category in the model. As already indicated, the value of information emerged as the core category and a driving force in the model. However, in some situations, the constraints overwhelmed the value of information, and the moderators had to act as intermediaries to overcome or reduce some of the constraints. Finally, a user-oriented model that attempted to highlight a concrete picture of rural women and health workers' information processes emerged, as Figs. 4.1—4.4 illustrate.

As many previous information researchers hardly focused on information use and what meaning information users gave to the information they accessed; therefore, the model and the rest of the findings presented in this book are important because they have highlighted how the interviewees interpreted and drew meaning from the information they accessed and used. The interviewees reported the difference that information had made in their lives and professional work when they accessed and used it, but they also narrated the frustrations and consequences of failure to access and use the needed information. The findings tend to orient the work towards the effect approach to information. As highlighted in Chapter Three (value of information), for example, once used, information had the capacity to make some effect on the user as well as on her/his acquaintances. According to Wersig (1997: 222), "the effect approach states that information is a specific effect of a specific process, usually on the part of the recipients in a communication process."

The effect or value of information was experienced, perceived and reported by the interviewees as having made a difference in their lives and/or professional activities. Interviewees were also able to create meaning from the difference that information made to them. Similarly, the effect and meaning of information reported in the findings were constructed by the interviewees and became a social reality. It was also noted that the meaning of information as reported by the interviewees and its subjectivity agreed with the mass communication Uses and Gratification theory (Watson, 2008), which has "another basic tenet...that the meaning of media experience can be learned only from people themselves. It is essentially subjective" (McQuail, 1994: 318). Furthermore, Jere and Davis (2011) recommended that future studies should investigate the uses and gratifications from a gender perspective.

Above all, the meaning that the interviewees garnered from the information they received through interactions with others yields a symbolic interactionist perspective, as articulated by Blumer (1969) and later extended by Snow (2001). Blumer (1969: 2) highlighted three major premises fundamental to symbolic interactionism: human beings act toward things[1] on the basis of the meanings that the things have for them; the meaning of such things is derived from, or arises out of, the social interaction that one has with one's fellows; these meanings are handled in, and modified through, an interpretative process used by the person when dealing with the things she/he encounters. Symbolic interactionism perceives meaning as arising in the process of interaction between people. According to Giddens (1997, 2013), a symbol is something that stands for something else. Symbolic interactionists reason that virtually all interactions between human individuals involve an exchange of symbols. Attention is directed to the detail of interpersonal interaction and how that detail is used to make sense of what others say and do. From the sociological point of view, the focus is usually on face-to-face interaction in the contexts of everyday life. Symbolic interactionism has been criticized for concentrating on small-scale rather than large-scale structures and processes. The major difference between symbolic interactionism and structural perspectives, for example, is that while symbolic interactionism concentrates on face-to-face contexts of social life (micro-social features), structuralism focuses mainly on cultural features of social activity (macro-social features); they both sprung from language but were developed with a different focus.

With that distinction, therefore, the findings tend to support symbolic interactionist perspectives. This does not mean that the cultural aspects that underpin women as care providers in the family and the culture embodied in social practice and activities such as burials or weddings (which were sources of information) are being ignored. What emerged from the findings commonly highlighted a face-to-face interaction either as a preferred source of information (eg, friends or colleagues) or as moderators. Interpersonal interactions emerged as important moderators of constraints to

---

[1]According to Blumer (1969), such things include everything that the human being may note in her/his world for example: physical objects such as books or trees; other human beings such as colleagues, friends, or relatives; institutions such as health units, church, or Government; guiding ideals such as individual independence or honesty; activities of others such as their requests and situations such as those an individual encounters in her/his daily life.

information access for both the women and health workers. Furthermore, whatever information was accessed, there was interaction with information sources and, hence, the Interaction-Value model that emerged from the findings.

The general findings, however, differ slightly from symbolic interactionism. First, the use of printed, electronic and audio information sources did not involve direct interaction between people, although it fits in the first premise of acting toward things. Second, when one considers sociological issues such as human action and social structure, one finds that symbolic interactionism emphasizes the creative and active components of human behavior as controlling the conditions of human lives, but it gives less attention to the constraining nature of social influences on the actions or activities of human beings. The findings vividly highlighted that factors such as gender or religious practices and values interfered in both women's and health workers' activities, especially on topics such as contraceptives and AIDS control. Hence, even though women interacted with health workers, accessed information about contraceptives and derived meaning out of the information they accessed during the interactions, some of them reported that they were not able to use that information due to the social forces around them. On the other hand, social factors were reported to have moderated information access for the interviewees in several ways.

The meaning of information for the women and health workers, the factors, and the behaviors were highlighted in the book. When the women, for example, perform their roles as caregivers or leaders and follow through with the commitments they entered into, they fulfill obligations that are defined by culture, local Government (in the case of local council (LC) leaders) and groups (in the case of group leaders). The information they access in the process, the meaning of that information to women and the various dissemination activities that proceed after interpreting and attributing a value to information have been clearly demonstrated in this book.

The interpretation of information and the interactions were also highlighted in Chapter Three (Actions in the value of information category). The women accessed some information, used it and disseminated it to others formally (eg, in meetings, and informally to relatives, friends and the community (interaction with networks). This agrees with the Two-step flow of communication model in a number of ways, for example, the women interviewees were opinion leaders who reported, among other things, that radio was the best channel from where health information was accessed in their rural settings. Second, the women leaders generally disseminated information to people of similar socio-economic status, but the people they disseminated information to had great respect for them and regarded them as knowledgeable. During the interviews, for example, women quoted the LC secretary for women as a source of various types of information. Hence, information moved from the source to the women who interpreted it and passed it on to other people who were reported to have been influenced in many ways; for example, some overcame misconceptions about family planning and other beliefs. Such interactions and personal influence have been reported in mass media literature (particularly under the Two-step flow) as playing a major role in spreading innovations and bringing about technical and cultural change. Furthermore, the interpretation of information sources, such as the media, is that they are more likely to provide information and raise awareness than to shape opinion or effect behavioral change, the media's influence depends on personal and social characteristics, and it is not always direct; instead, the media may first influence opinion leaders, who then influence other people (Defleur and Dennis, 1989); therefore, interpersonal communication is often necessary to bring about the desired behavioral change (Paek et al., 2008).

In addition, the findings showed that information access (and use) may involve only one step. This was when both women and health workers reported having accessed information directly from, for example, the media and it had an impact on them. However, there were many situations in which the information process involved several stages. For example, religious leaders accessed information and announced it in church; those who were in church received and passed the information on to friends and relatives or to LC/women group meetings, who also passed it on to others. Similarly, health workers reported that they accessed information from various sources and then passed it on to other health workers or women in seminars/documents they produced, and these people then shared it with others. Such findings have been reported to have led to the evolution of the two-step flow into a multistep flow model (Watson, 2008).

Studies conducted elsewhere in rural Africa had similar findings; for example, Bosompra (1989: 1138) reported that "Our findings underscored the relevance of the two-step and multi-step flow models of the communication process to health information dissemination in Ghana. Respondents relied almost equally on conversation with family and friends on the one hand and radio on the other, for information on the selected health topics. Health workers and a traditional communication channel... also played significant roles."

It has been pointed out in mass communication literature that in the two-step or multi- step flow models, individuals are not social isolates, but members of social groups interacting with other people. Furthermore, that response and reaction to a media message may not be direct and immediate, but may be mediated through and influenced by these social relationships (McQuail and Windahl, 1981; McQuail, 1994, 2010).

These findings inevitably lead the discussion to the traditional "social networks," which emerged as important access and use factors in this work. Despite their contribution to information access and use, however, social networks had not received due attention by information studies researchers, as Belderson (1999) reported that "the influence of the social network as a source of information remains a comparatively neglected area. A number of publications has devoted some discussion to the influence of family and friends... but studies with a primary focus on this source are lacking. On some occasions, this source has even been deliberately overlooked (or) excluded" (p. 229).

While generally agreeing with the statement by Belderson (1999), Dubnjakovic (2015) pointed out that, on the contrary, the social network analysis (SNA) has been used by many researchers in the social and physical sciences for a long time. However, by early 2015, only two studies by Haythornthwaite (1996) and an update by Shultz-Jones (2009) were identified in the major information science journals. Consequently, Dubnjakovic (2015) emphasized the importance of SNA in information behavior research and how cohesion and brokerage as characteristics of SNA and reciprocity, popularity and attributes such as gender are important parameters.

In this book, social networks enhanced the information process by moderating constraints to information access and use as indicated in Chapter Two (Sections 2.3.4, 2.4.4 and 2.5.4, Moderators). The findings, therefore, agreed with those of Belderson (1999), who pointed out that the informal network was a widely accessible and utilizable source that was well suited to the delivery of information without a high degree of active seeking and in a natural and unpatronizing environment. Furthermore, Weiten et al. (1991: 83) pointed out that "friends may be good for your health! This startling conclusion emerges from studies on social support... it involves various types of aid and succor provided by members of one's social networks... Social support serves four important functions... (one of which is): Information support involves providing advice... This includes discussing possible solutions and the relative merits."

The findings on which the Interaction-Value model is based are, therefore, consistent with the SNA by Haythornthwaite (1996), who pointed out that informal information exchange routes develop based on local needs and that "Information is made useful... by being forwarded to others... The need to exchange or receive information can lead to the establishment of new information routes. Network analysis describes these routes and the course of information... not just delivery from supplier to client, but also from client to relative, to subordinate, to superior, to friend, or to acquaintance, and from there to others" (p. 339).

Furthermore, Cooke (2015) indicated that SNA started much earlier than social media and focuses on the relationships between entities. It highlights the differences between formal and informal structures as well as the nodes or actors. SNA also shows the impact on information and knowledge flows and identifies bottlenecks, information brokers and reciprocity, as the findings presented in this book have done.

Related to the interaction and social networks is the information behavior. As already pointed out, the findings showed that in the case of women, passive access was the principle behavioral mode of health information acquisition. Conversations with relatives, friends and village-mates provided a great deal of information. Most women also listened to the radio and preachers in the church or the mosque, received text messages on mobile phones and watched drama, films and television. Active health information seeking was mainly reported in critical incidents or situations, either personal or family, and as women leaders when they were required to attend seminars or to collect and disseminate information to their communities. The women's findings differ from those of some previous works such as "Wilson's formulation of the information seeking processes implicitly takes active searching as the principle mode as does Ellis's behavioural model of information seeking" (Wilson, 1997: 562).

The findings, however, agree with those of Belderson (1999). She noted that a great deal of information research had concentrated on active seeking in which the search for information involves a conscious attempt to satisfy a predefined information need. "Information behaviour in the context of the present study was very different - participants tended to be confronted with information rather than actively seeking it. Acquisition of nutrition information ranged from purely accidental (e.g. while watching a television programme) to semi-purposeful (e.g. consulting a doctor about a health problem and receiving nutrition information as a consequence)... Passive acquisition of information is not so directly observable and has received relatively little research attention" (p. 233).

Passive access to information does not seem to have attracted much attention from information researchers. This is probably because it is considered inappropriate because it tends not to have direct relevance to information systems design, which most information studies target. Passive access, however, emerged inductively from the data, and it could not be ignored. The dynamic nature of women's information behavior was also an important finding. The fact that information accessed passively sparked various active behavior either by disseminating the information to others or by actively seeking more information to confirm or clarify what had initially been received is something worth documenting. Hence, although the women mostly accessed information passively, their subsequent information behavior was active.

However, health workers actively sought most of the information. Health workers' information needs and their information seeking behavior in this book generally agree with Wilson's (1997) model of information behavior already mentioned. The findings also support Dervin's sense-making theory to some extent. The theory states that people seek information when they have identified gaps in their knowledge that prevent them from making sense of a situation in which they find themselves, from solving a problem at hand, or from making an informed decision (Dervin, 1992). That theory can therefore explain generally when and why most health workers seek information. The findings presented in this book, however, have gone a step further to highlight the subsequent information activities, which impacted on a number of constraints, thereby enabling other people to access and use information. Furthermore, the two models (Wilson's and Dervin's) may not explain, to the same extent, the information behavior and information needs of rural women in Uganda.

Many women reported that they had "gaps in their knowledge" about certain diseases or health problems. Dervin's theory (1992) would suggest that they seek information about these issues to make sense of the situation. However, women went on to report that they did not seek information in the majority of the situations, thereby making these needs latent rather than active. Such gaps in the women's knowledge remained until they were exposed to information passively, as already reported. Wilson (1997) acknowledged the fact that people do not always seek information when they have knowledge gaps: "The fact that purely cognitive drives cannot explain information seeking behaviour is attested to by the fact that, even in critical circumstances when the gaps in their knowledge are evident, people do not always seek medical information" (p. 555).

The above observation was also confirmed by Palsdottir (2007, 2010) in a study of health and lifestyle information seeking in which it was reported that information encountering was an integral feature of information behavior. Information was encountered more often than sought on purpose by all clusters. Clusters that were active in purposive information seeking were also active in information encountering.

Women's information behavior in this book was also different from the blunters in the monitoring and blunting theory (Baker, 1995, 2005), because the blunters deliberately avoid getting information. In fact, when asked whether they would prefer getting information or not if they were faced with a life-threatening problem, all women interviewees, except two, answered "yes."

The traditional social network and the Multi-step flow model, discussed previously, may, to a certain extent, explain women's information behavior better than other models/theories. However, the finding showed that women's access to and use of information, as well as their information behavior, could not be fully explained by isolated or single factors; it was a complex and intricate process that involved a number of factors and actors and depended on, to a large extent, the situation within the family, community and district or country in general, as well as personal attributes. It is, therefore, a multivariate-based and quite dynamic situation.

Thus, both passive and active behavior were prominent in the Interaction-Value model. The information behavior in this book can therefore be summed up as access with or without seeking.

Among other things, the Interaction-Value model differs from the models discussed previously in that the value of information, more than anything else, was the driver in the various information activities that resulted in the model.

Hence, the Interaction-Value model cannot be fully explained by purely information concepts. It is a multidisciplinary model (like several others in information science such as Wilson's, 1997) with orientation to and concepts from information, communication, behavioral science, sociology or social psychology, and (to some extent) gender and feminist approaches.

To evaluate the use and saturation of the Interaction-Value model, other studies in related areas were examined as elaborated in this chapter. Furthermore, a number of information studies that have used the Grounded theory approach were reviewed. This approach enables researchers to build accurate or naturalistic models in information and other fields because of its ability to allow models to emerge from real-world situations. In this book, the Interaction-Value model was derived from the experiences, perceptions and meaning that women and health workers had of their information environment and the role of information in their everyday lives and activities.

The next section presents the strengths and limitations of the model.

4.5 STRENGTHS AND LIMITATIONS OF THE MODEL

The section has two subsections: strengths and limitations.

4.5.1 STRENGTHS

The strengths of the Interaction-Value model include:

i. The model demonstrates its suitability for the phenomenon of information access and use (as processes that depend on the value of information) in its attention to the identification of information sources and needs, the various constraints to information access and use, and the factors or structures that moderate the constraints to enable people to access and use health information. The model, therefore, consists of related categories that can be used to explain the phenomenon of health information access and use in rural Uganda.

ii. The model is understandable because it is made of concepts that are inherent in women's and health workers' everyday lives and work, which highlight the realities of an information environment in rural Uganda.

iii. It is applicable or transferable because it is a general model that can operate in a variety of information contexts and cultures in Africa, other developing countries and some communities in the developed world. Hence, it is flexible enough to manage change and situational realities.

iv. The model has potential for linkages to related information access and use phenomena or processes, such as those in agriculture and related biological science fields.

v. The model provides some means of control of the information process by indicating the necessary components of information activities in everyday situations, as well as the moderators that regulate or intercept the constraints to information access and use. An information provider, for example, would be able to judge what information to provide in what situations, and would also be able to change the information strategy as the situation changes.

The strengths of the Interaction-Value model outlined above are in line with the recommendations made by Glaser and Strauss (1967) that the practical application of the Grounded theory, whether substantive or formal, requires developing a theory with at least four interrelated properties: "the theory must closely fit the substantive area in which it will be used. Second, it must be readily understandable by laymen concerned with this area. Third, it must be sufficiently general to be applicable to a multitude of diverse daily situations within the substantive area, not to just a specific type of situation. Fourth, it must allow the user partial control over the structure and process of daily situations as they change through time" (p. 237).

4.5.2 LIMITATIONS

Some reflection on the limitations of using a holistic inductive approach with a Grounded theory in a Ugandan setting is presented. It was noted that the methodological paradigm used in the study had some limitations arising mainly from the lack of a common language in the study areas, cultural taboos and the research fatigue expressed by some interviewees. Each of the issues is briefly discussed in this section.

i. *Language/Vernacular*

The Grounded theory requires grounding in data as it depends on rich qualitative data. A situation, when interviews were conducted not only in the vernacular but also in four languages (three Bantu and one Luo) and later translated the data in English, tended to create gaps and lose some amount of information. This may have affected the quality of the data and the momentum for formulation of concepts on which the theory is founded. The different languages or lack of a national/common language would also make it difficult to use a theoretical sampling strategy in different rural areas. Patton (1990) made the following observations about language differences: "The data from interviews are words. It is tricky enough to be sure what a person means when using a common language, but words can take on a very different meaning in other cultures. . . The situation becomes more precarious when a translator or interpreter must be used because of language differences. . . Interpreters often want to be helpful by summarising and explaining responses. This contaminates the interviewee's actual response with the interpreter's explanation to such an extent that you can no longer be sure whose perceptions you have - the interpreter or the interviewee. There are also words and ideas that simply can't be translated. . . Patton (1990) has published a whole book on untranslatable words with special meanings in other cultures" (pp. 338–339).

The problem of words or phrases "that can't simply be translated" made the researcher/ author drop some interesting proverbs that would have acted as indigenous concepts in the analysis (catchy, suggesting and summarizing words or phrases used by interviewees). This was a problem mainly with the data from the women interviewees; for example, a proverb such as "ekiija ninkimanya" (in Runyankole–Bushenyi) presented in Chapter Two (Section 2.3, coping and information behavior) does not really translate to "forewarned is forearmed," although this is the closest English translation. Similarly, "tokamanya" generally refers to "you never know," which was used to describe a constraint to information use in Chapter Two (Section 2.3.3). The analyst/author had to find an inductive term and referred to it as "attitudes/views" held by people that made them fail to use the information they had received (eg, about the immunization program). Hence, instead of catchy phrases such as proverbs, the analyst/author used "in vivo codes," which were words or sentences from the interviews that were easier and/or more straight forward to translate from the vernacular to English without losing the "value-added" meaning, as in the case of proverbs. In most cases, however, the analyst/author had to find inductive terms or concepts.

Translating all interviews from the different languages to English was laborious and time-consuming. Many times, the analyst/author had to go back to the original field notes/recording to cross-check the translated versions to make sure that the concept she was trying to come up with really fits the original data.

ii. *Cultural taboos*

According to the societal norms, Ugandan women do not discuss sex and related issues with strangers. This becomes even more problematic when the opposite sex is involved. The two male research assistants reported, for example, that topics such as condom use and Sexually Transmitted Infections (STIs) were not easily talked about; therefore, it was difficult to probe and get in-depth information about these topics. This could have affected the quality of data collected for these and related topics from some traditional birth attendants who were interviewed by the male research assistants at the beginning of the data collection period. However, the problem was identified early and the author joined the male research assistants in the interview sessions. Given the fact that the majority of women were interviewed by female research assistants, this problem did not greatly impinge on the overall reliability of the data on sex-related topics. This experience, however, may be useful to future researchers.

iii. *Research fatigue*

Many research studies and/or projects have been conducted in different parts of Uganda. Several interviewees raised some concerns about being the subject of research with hardly any tangible personal or professional benefits accruing at the end of the exercise. Others pointed out that they neither get to know the results nor get to see the impact of research projects in which they participate. This could also have affected the quality of data collected from such individuals.

However, to get the cooperation of the interviewees, the researcher explained that this was an academic study, which made them receptive. Furthermore, to increase their willingness to participate, interviewees had the freedom to choose the interview time and venue convenient to them. The majority of health workers, for example, preferred the evenings (after duty) or lunch time during the weekends, and the venues were either the health units or somewhere else that was quiet. When needed, refreshments and snacks were provided by the health workers and/or the author during the interviews. The author/researcher distributed some back issues of the Uganda health information digest and other health-related periodicals to health workers who needed them. Interviewees were, therefore, able to gain something tangible from their participation in the research. Although this issue may not have greatly affected the quality of data, the general research fatigue in rural areas is something that future researchers need to take into consideration.

Despite the identified limitations, the findings, as Chapter Two demonstrated, were very rich and managed to provide what Glaser (1978) referred to as "thick description." This shall advance our knowledge and understanding of the information access and use in rural Uganda, which can be applied in other parts of the world.

Finally, given the rich nature of the findings already presented and a new model of information behavior, an overview of the conclusions is provided and the implications are highlighted in the next, and last, chapter of the book.

REFERENCES

Apalayine, G.B., Ehikhamenor, F.A., 1996. The information needs and sources of Primary health care workers in the Upper East Region of Ghana. Journal of Information Science 22 (5), 367–373.

Baker, L.M., 1995. A new method for studying patients information needs and information seeking patterns. In: Lloyd-Williams, M. (Ed.), Health Information Management Research. Proceedings of the First International Symposium. University of Sheffield, Department of Information Studies, Centre for Health Information Management Research, Sheffield, pp. 67–75, April 5–7.

Baker, L.M., 2005. Monitoring and blunting. In: Fisher, K., et al., (Eds.), Theories of Information Behaviour. American Society for Information Science and Technology, Medford, NJ.

Belderson, P., 1999. Food Choice in Older Adults: The Role of Nutrition Information. PhD. Thesis, University of Sheffield.

Blumer, H., 1969. Symbolic Interactionism: Perspective and Method. Prentice-Hall, Englewood Cliffs, NJ.

Bonfadelli, H., 2002. The Internet and knowledge gaps: a theoretical and empirical investigation. Eur. J. Commun. 17 (1), 65–84.

Bosompra, K., 1989. Dissemination of health information among rural dwellers in Africa: a Ghanaian experience. Soc. Sci. Med. 29 (9), 1133–1140.

Bryman, A., Burgess, R., 1994. Reflections on qualitative data analysis. In: Bryman, A., Burgess, R. (Eds.), Analysing Qualitative Data. Routledge, London.

Cooke, L., 2015. Social network analysis. EAHIL conference. https://eahil2015.wordpress.com/workshop-presentations (accessed 21.09.15).

Defleur, M.L., Dennis, E.E., 1989. Understanding Mass Communication. third ed. Houghton Mifflin, Boston, MA.

Dervin, B., 1992. From the mind's eye of the user: the sense-making qualitative-quantitative methodology. In: Glazier, J.D., Powell, R. (Eds.), Qualitative Research in Information Management. Libraries Unlimited, Englewood, CO, pp. 61–84.

Dervin, B., Nilan, M., 1986. Information needs and uses. Annu. Rev. Inf. Sci. Technol. 21, 3–33.

Dervin, B., et al., 2006. Beyond communication: research as communicating. Making user and audience studies matter- paper 2. Inf. Res. 12 (1).

Dubnjakovic, A., 2015. Social network analysis in information behaviour research: a case for exponential random graph models. Paper Presented at the Qualitative and Quantitative Methods in Libraries (QQML) Conference, Paris, May 2015. http://www.isast.org/images/FINAL_e-book_version.pdf; http://www.qqml.net/journal_issues.html.

Ellis, D., 1993. Modelling the information seeking patterns of academic researchers: a Grounded theory approach. Libr. Q. 63 (4), 469–486.

Eveland, W.P., Scheufele, D.A., 2000. Connecting news media use with gaps in knowledge and participation. Polit. Commun. 17 (3), 215–237.

Giddens, A., 1997. Sociological Theory. third ed. Blackwell Publishers, Oxford.

Giddens, A., 2013. The Consequences of Modernity. Polity Press, Cambridge.

Glaser, B., 1978. Theoretical Sensitivity. Sociology Press, Mill Valley, CA.

Glaser, B., Strauss, A., 1967. The Discovery of Grounded Theory: Strategies for Qualitative Research. Aldine de Gruyter, New York, NY.

Guttman, N., Salmon, C.T., 2004. Guilt, fear, stigma and knowledge gaps: ethical issues in public health communication interventions. Bioethics 18 (6), 531–552.

Haythornthwaite, C., 1996. Social network analysis: an approach and technique for the study of information exchange. Libr. Inf. Sci. Res. 18, 323–342.

Ilie, V., et al., 2005. Gender differences in perceptions and use of communication. Inf. Resour. Manage. J. 18 (3), 13–31.

Jere, M., Davis, S., 2011. An application of uses and gratifications theory to compare consumer motivations for magazine and Internet usage among South African women's magazine readers. South. Afr. Bus. Rev. 15 (1), 1–27.

Johnson, J., Meischke, H., 1991. Women's preferences for cancer information from specific communication channels. Am. Behav. Sci. 34, 742–755.

Kapiriri, L., Bondy, S.J., 2006. Health practitioners' and health planners' information needs and seeking behavior for decision making in Uganda. Int. J. Med. Inform. 75 (10–11), 714–721.

Kuhlthau, C., 1993. A principle of uncertainty for information seeking. J. Doc. 49 (4), 339–355.

McQuail, D., 1994. Mass Communication Theory: An Introduction. third ed. Sage, London.

McQuail, D., 2010. McQuail's Reader in Mass Communication Theory. sixth ed. Sage, London.

McQuail, D., Windahl, S., 1981. Communication Models. Longman, New York, NY.

Mutua, E., 1997. How rural women communicate in Kenya. Focus Int. Comp. Libr. 28 (2), 73—74.

Narayan, U., 1989. The project of feminist epistemology: perspectives from a non-Western feminist. In: Jaggar, A., Bordo, S. (Eds.), Gender/Body/Knowledge: Feminist Reconstructions of Being and Knowing. Rutgers University, New Brunswick NJ.

Paek, H., et al., 2008. The contextual effects of gender norms, communication and social capital on family planning behaviors in Uganda: a multilevel approach. Health Educ. Behav. 35 (4), 461, http://heb.sagepub.com/cgi/content/abstract/35/4/461.

Palsdottir, A., 2007. Patterns of information seeking behaviour: the relationship between purposive information seeking and information encountering. In: Bath, P., et al., (Eds.), Proceedings of the 12th International Symposium for Health Information Management Research (ISHMR). University of Sheffield, Sheffield, pp. 3—15.

Palsdottir, A., 2010. The connection between purposive information seeking and information encountering: a study of Icelanders' health and lifestyle information seeking. J. Doc. 66 (2), 224—244.

Patrikios, H., 1985. Socio-political changes in developing countries: the concerns of the medical librarian, In: Medical Libraries, One World: Resources, Co-Operation, Services. Proceedings of the Fifth International Congress on Medical Librarianship, Tokyo, Japan.

Patton, M.Q., 1990. Qualitative Evaluation and Research Methods. Sage Publications, London.

Saracevic, T., Kantor, P., 1997. Studying the value of library and information services. J. Am. Soc. Inf. Sci. 48 (6), 527—542.

Shore, E. 2015. Measures of our time. http://www.ifla.org/node/9847; https://becker.wustl.edu/impact-assessment/how-to-use.

Shultz-Jones, B., 2009. Examining information behaviour through social networks: an interdisciplinary review. J. Doc. 65 (4), 592—631.

Snow, D., 2001. Extending and broadening Blumer's conceptualisation of symbolic interactionism. Symb. Interact. 24 (3), 367—377.

Strauss, A., Corbin, J., 1990. Basics of Qualitative Research: Grounded Theory Procedures and Techniques. Sage Publications, Newbury Park, CA.

Talja, S., et al., 1999. The production of context in information seeking research: a metatheoretical view. Inf. Process. Manage. 35, 751—763.

Turner, B., Martin, P., 1986. Grounded theory and organisational research. J. Appl. Behav. Sci. 22 (2), 141—157.

Wallman, S., 1996. Kampala Women Getting By. James Currey, London.

Watson, J., 2008. Media Communication: An Introduction to Theory and Process. third ed. Palgrave Macmillan, Hampshire.

Weiten, W., et al., 1991. *Psychology Applied to Modern Life:* Adjustment *in the 90's*. third ed. Brooks/Cole Publishing company, Pacific Grove, CA.

Wersig, G., 1997. Information theory. In: Feather, J., Sturges, P. (Eds.), International Encyclopaedia of Information and Library Science. Routledge, London.

Wilson, T., 1981. On user studies and information needs. J. Doc. 37 (1), 3—15.

Wilson, T., 1997. Information behaviour: an interdisciplinary perspective. Inf. Process. Manage. 33 (4), 551—572.

IMPLICATIONS FOR THEORY, PRACTICE AND THE FUTURE

5.1 INTRODUCTION

The author concludes the work with two quotes from a woman and a health worker interviewed:

> Although health problems are many, I have been able to solve some using the information I have received; for example, I have managed to control malaria in my family by keeping our home free from mosquito breeding sites, and we all sleep under 'mosquito' nets. ...in my view, I think that improving the provision of health information to women would be the beginning of a better life for a rural community because women provide care for everybody in the family
>
> **A woman from Bushenyi district, Western Uganda**

> we changed the management of post-operative infections after reading evidence based literature and applying it ... and this has reduced the wastage of medicine and the burden to the patients
>
> **A doctor from Masaka district, Central Uganda**

The quotes wrap up the work quite well. First, they have confirmed that when people are informed, they are healthy; and when health workers apply new knowledge in their work, there are many benefits to health care. The quotes have, therefore, highlighted the value attributed to information in the betterment of health. That value has been demonstrated in the book to have driven and sustained the various information processes. Second, the quotes have also confirmed that the main concern of the book was to understand the issues about access and use of information in rural Uganda, and how women and health workers perceived and interpreted these issues. Third, the first quote has demonstrated that in an African family, women nurse and care for the sick; therefore, targeting women in information provision would have a multiplier effect. Furthermore, the second quote emphasizes the importance of health workers' access to and use of evidence-based literature and the impact it has on health outcomes. Finally, given all the benefits, there is a need to improve and sustain the health information provision at all levels of Primary Health Care (PHC).

Women's role as health care providers in the family was clearly demonstrated in the book. What made the findings so distinctly African was the extended family aspects. Women needed information to enable them to provide care for both their nuclear and extended family members. For example, when women accessed and used information, the value of that information was reported in relation to its effect or benefit to the nuclear and/or extended family members.

The book has demonstrated that Uganda, as a Sub-Saharan African country, is struggling to ensure that the needed information is accessed and used by people to prevent the many diseases

Informed and Healthy. DOI: http://dx.doi.org/10.1016/B978-0-12-804290-8.00005-1

that are preventable and to equip the health workers with appropriate information to make informed and evidence-based clinical decisions that will impact the health care outcomes. Hence, access and use of health information are the basic issues in low- and middle-income countries and among the underserved communities living in developed countries.

5.2 IMPLICATIONS FOR THEORY AND APPLICABILITY OF FINDINGS

The book has presented findings from rural people who are usually marginalized or left out of high-level information studies research, which has tended to focus on academics and professionals. For example, Ellis (1993) focused on academic researchers, Palmer (1990) focused on agriculturists, Soto (1992) focused on dental professionals, Odini (1995) studied engineers, Choo and Marton (2003) focused on women in the IT professions, Mabawonku (2006) focused on women in Nigeria's public service, Namaru et al. (2009) focused on undergraduate university students, and Mbondji et al. (2014) focused on policy makers and data managers in national ministries of health, national statistics offices, health programs, donors and technical agencies in Africa. In Uganda, Kapiriri and Bondy (2006) focused on higher-level health practitioners, planners and policy makers at district and national levels. Kinengyere and Olander (2011) focused on medical academics. Lugya and Mbawaki (2011) and Musoke et al. (2014) focused on academic library users, and there were several unpublished theses and dissertations, in Ugandan University libraries, which focused on information technology, women, health, etc. Some literature about women or health workers was mainly descriptive rather than interpretative. As pointed out in Chapter One, focusing research on rural health workers is an important step in improving their information environment. This would also enhance the information provision to the communities they serve because the majority of Ugandans live in rural areas.

The book also demonstrated that lay people can access a significant amount of information without active seeking. This was particularly true for the women. The book further showed that information access was a phenomena resulting from the interaction between individuals and information sources in the context of life-related situations (in the case of women) and mainly, but not exclusively, work-related situations (in the case of health workers) that provoked information needs, information use and information behavior, with the value of information at the center of these activities.

Furthermore, the book has addressed several gaps highlighted by Palsdottir (2007), for example, that most previous studies examined active and passive information behavior separately. The approach of this book is holistic, because the different styles of information behavior are addressed. Chapter Two of the book has, among other things, demonstrated the dynamic nature of information behavior where, in one incident, for example, the information behavior of an individual changed from passive to active and vice versa. The book has also bridged another gap identified by the same author that "an improved knowledge of information seeking behaviour, and how the different groups within society can be reached in a more effective way, is therefore of importance for health promotion and public health practice, as well as of theoretical value" (p. 3). The moderators of constraints to information access and use presented in Chapters Two and Four, and the value attributed to information and its effects elaborated in Chapters Three and Four have also addressed these concerns.

This book has also attempted to address a need reported by Ford (2015) that the understanding of information behavior may help maximize the quality and effectiveness of the way information is presented, sought, discovered and used.

Furthermore, the book seems to acknowledge a symbolic interactionist perspective because it focuses on the importance of the meanings that emerged as people defined their information situations through interpersonal interactions. Such interactions were highlighted in all of the categories of the Interaction-Value model presented in Chapter Four. However, symbolic interactionism stresses the creative and active aspects of human behavior and tends to give less attention to the social factors that may be beyond the control of human beings but nevertheless constrain their actions. This book concludes that both issues are important to the information access and use in rural Uganda. The various face-to-face interactions moderated information access and use when they took place, but there were situations that had not been moderated by the interactions, and the social factors overwhelmed the information process.

The book has, therefore, succeeded in giving an in-depth view of access to and use of information by women and health workers in rural Uganda. We have progressed from knowing about the sources of information to what that information actually does once it is accessed and used, and how valuable it is to the professional activities of health workers and the lives of women and those under their care. The value of information elaborated on in the book and the subsequent actions have shown that a successful information activity has great impact on health care. Although it is not strictly possible to generalize the findings from this purposive sample to all rural areas, it is likely that the issues identified in the book can apply to other lower levels of PHC in rural Uganda, as well as to other low- and middle-income countries.

Besides the lack of a national language in Uganda, the methodological approach used in the study, on which the book is based, proved meaningful for future applications in related studies in several ways, for example:

i. *Sampling* health workers and women at the lower levels of PHC allowed a combination of professionals, apprentices and nonprofessionals to be studied to gain a qualitative view of their information environment. Among the health workers, the sample allowed a combination of disciplines (medicine, nursing, midwifery, as well as apprenticeship in the case of Traditional Birth Attendants (TBAs)) with job patterns (health unit-based/private practice and community workers). Among the women, the sample brought in different categories of leaders and educational levels.

ii. A semi-structured but open-ended interview *method* provided rich insight into the factors affecting the interviewees' access and use of health information. It was more appropriate than an interview guide because it reduced variations among the different interviewers, and it made it possible for the interviewees to answer the same questions, which in turn facilitated cross-case data analysis.

Furthermore, the inclusion of the description of critical incidents greatly enriched the data because it brought in a critical aspect that highlighted the information needs, sources and behavior in situations that were different from the ordinary. Women, for example, who were generally passive exhibited active information seeking behavior in critical situations.

iii. A holistic inductive approach with a Grounded theory *analysis* can be used to study the information behavior in general and access to and use of information by women, health workers or other professionals in Africa and other parts of the world.

With regard to whether the Interaction-Value model can be applied, replicated, or transferred elsewhere, Chapter Four (Section 4.5.1) highlighted the model's ability to be replicated as one of its strength. Furthermore, Chapter Four (Section 4.4) discussed the Interaction-Value model and related it to existing models to give it a wider view in the field of information science and an international appeal. The Interaction-Value model is indeed an information science model that can generally be applied in situations where people access and use information. Although the study, on which the model is based, was conducted in a Sub-Saharan African setting, the findings have international transferability. This was demonstrated, for example, by Smith (2002), who reported that the interactivity in the Interaction-Value model was added to the Slawson and Shaughnessy formula. The formula was $U = \dfrac{R \times V}{W}$.

Where U = usefulness of the information to health workers, R = relevance of the information, V = validity of the information and W = work to access the information. In other words, the most useful information for health workers is information that is relevant to their practice, is valid, and does not take too much work to access. "After listening to a presentation by Maria Musoke, a researcher from Uganda, on the usefulness of information to rural health workers in Uganda, I added 'interactivity' to the top line of the equation. The information is still more useful if you can interact with the source and interrogate it. The formula provides a test of the ways in which doctors look for information they need" (editorial, p. 2). This contribution to theory and applicability was published by the British Medical Journal.

From a theoretical point of view, therefore, the book has provided both the qualitative findings and an information model on a rarely researched PHC level. The findings shall complement the quantitative and sometimes descriptive reports presented in previous studies. Several information professionals had already asked the author to use some of her field quotes in their information project proposals. This was in particular reference to the quotes about constraints to information use (see Chapter Two) and value of information (see Chapter Three). This shows, among other things, that there had not been such qualitative findings previously that could be used to blend or enrich the proposals, which are based on and justified by quantitative data. The book has therefore made a contribution to information research.

5.3 IMPLICATIONS FOR HEALTH INFORMATICS AND INFORMATION PROVISION IN GENERAL

Although the book is based on an academic piece of work, it is more applied than basic. Hence, at a practical level, the book has various implications for information provision. Besides advancing our knowledge, qualitative research provides the necessary details needed to design systems and improve service delivery. Furthermore, access to health information has become topical, making the book relevant to local and international agencies.

Among other things, the book used a holistic approach, which has made it possible to elaborate on the women's and health workers' information activities. This can guide the improvement of health information provision in Uganda and other low-income countries. It is therefore important to address the identified constraints as well as strengthen the existing (and potential) moderators and information sources that the book has demonstrated to be effective in supporting health information provision to rural areas. Some examples are highlighted below.

A regular rural outreach health information literacy program is recommended. This initiative was evaluated and found to be useful (Musoke, 2014), and it was reported by many interviewees in Chapter Two as having provided valuable information. Sessions were conducted during the outreach to demonstrate the available information resources that rural health workers and lay people could access directly or indirectly by making requests for document delivery.

To the developing world, the key issue was access and use of information, which led to development agencies such as the United Nations under Kofi Annan to convince the world's leading publishers to let the poor countries' health professionals access the current literature. Consequently, the Health InterNetwork Access to Research Initiative (HINARI) and other Research for Life resources were established for use by low-income countries such as Uganda. Among other things, the book has reported the progress and challenges experienced so far.

The increase in electronic availability of documents and the gradually improving speed of online access have particularly suited the information needs of health workers who typically require a timely response to urgent and sometimes emergency information needs. However, given the fact that some health workers had not fully benefitted from such information resources, there was a need to put in place interventions that would address the identified challenges. Otherwise, the globalization of information would threaten to disempower the developing world further! As the economy gradually improves, the information and communication technology (ICT) infrastructure in rural areas is likely to improve. Our responsibility as health information providers is to maximize the benefits of online information resources and minimize their weaknesses. For example, information providers who repackage online information in different formats to make it available to health workers, who would otherwise not access it, should be emulated and the sustainability of such activities should be supported.

The use of ICTs for health referred to as "eHealth" is an important initiative, and there are national eHealth associations formed in some African countries. However, the issue of how to leverage eHealth to advance health services, particularly in low-resource settings, remained a challenge (Rockefeller Foundation, 2008). Later, WHO (2010) highlighted the need for a global, long-term and collaborative approach so that all citizens of the world may benefit from the best eHealth solution possible.

Initiatives such as the Rockefeller Foundation support to Health Informatics Building Blocks (HIBBs) in Sub-Saharan Africa are commendable. The initiative focused on clinical and health informatics training in low-resource countries to enable better support of community care and public health services.

The World Information Society (2014) reported that the urban—rural digital divide prevailed in Uganda, but that it was closing with the availability of affordable mobile phone services in rural areas. This was similar to what had been presented in the book, which highlighted the use of mobile phones by women and health workers and the difference it had made in their health information activities.

While agreeing with the previous statement, Omaswa and Crisp (2014) pointed out that Africa and other developing regions, without the legacy of earlier infrastructure, are adopting many applications earlier than the more industrialized countries. They highlighted the use of "mobile phone technology to train the community health workers... Phones throughout Africa are used for tasks such as making clinic appointments and follow-ups, transmitting test results... email and videos support patient consultations and allow isolated clinicians to seek advice from colleagues..." (p. 317).

Organizations involved in successful health information activities using mobile technologies should, therefore, be emulated. For example, the Uganda Chartered Healthnet, among other things, had implemented several projects using mobile technologies to deliver current and relevant electronic information to health workers in rural Uganda. The use of mobile technologies is likely to enhance access to information in the future. Information providers should, therefore, make use of such avenues.

Previously, the Commission for Africa (2007) report recommended that the continent's focus should be on research information and communication. This was because scientific skills and knowledge enabled countries to find solutions to their own problems and brought about step-by-step changes in areas from health, water supply, sanitation and energy to the new challenges of urbanization and climate change, which unlock the potential of innovation and technology to accelerate economic growth and enter the global economy. Health information professionals who handle most of the research information need to contribute to the recommended changes and approaches.

Furthermore, at the end of the Millennium Development Goals, a new development agenda, the Sustainable Development Goals (SDGs) for 2016—2030 with approaches to improving people's lives was launched by the United Nations in September 2015. The book is very timely because it presents the voices of health workers and lay people highlighting the value of information and the effects it has had on health care. The book also highlights the need to address the challenging ICT infrastructure in low-income countries to be able to fully take advantage of the increased online resources that will improve people's lives and support sustainable development.

At a national level, Uganda's "Vision 2040" promised a better future as it planned to address the strategic bottlenecks that had constrained Uganda's socioeconomic development (The National Planning Authority, 2013). Furthermore, the National Planning Authority of Uganda had put in place strategies for implementing the SDGs. At an international level, and with the support from the International Federation of Library Associations (IFLA) and its Lyon Declaration (2014), librarians started documenting examples of how libraries had contributed to the success of SDGs.

The subsequent parts of this section outline some specific implications for the women, health workers and information professionals.

5.3.1 IMPLICATIONS FOR INFORMATION PROVISION TO WOMEN

A participatory and multi-sectoral approach involving front-line health workers, the local authority (LCs), the church and other faith-based organizations, women's groups and other development agencies is an example of effective provision of information to women. The detailed methods by which effective information services should be delivered can vary according to the local conditions, but the common principles remain. The provision of health information, particularly to rural areas, in a readily accessible and usable format should be an area of significant activity for health information providers, including the relevant sections of the Ministry of Health (MoH). More information needs to be simplified, translated in local languages, illustrated for those who are not able to read, and presented in electronic (such as text messages), audio (through radio), visual (television, DVDs, through drama or popular theater, print) and other formats (eg, seminars) as the findings have shown. The social networks that proved both valuable and effective as moderators and sources of information should be recognized and utilized more.

Health education and promotion focusing on specific health problem(s) of a particular area are crucial. The benefits of effective health education have been clearly demonstrated in the book. If "prevention is still better than cure," then there is no need to wait for cholera, Ebola or other epidemics. Furthermore, the results of the AIDS prevention campaigns, which holistically involved all stakeholders, have been well documented. Other killer diseases and health problems such as malaria, maternal illnesses, diabetes and cancer need such effective health education and promotion programs.

Information provided by health workers to patients was generally considered insufficient. The findings showed, for example, that although the majority of women interviewees preferred to get more information (monitors) about their long-term and life-threatening illnesses or the illnesses of those they nursed, two women reported that information would make them more worried and, hence, worsen their situation (blunters). It is, therefore, important that health workers assess the patients' orientation to information so that the monitors are provided with as much appropriate information as possible to assist them in coping with illnesses. The women interviewees generally reported that health workers provided too little information to patients and their carers, something that needs to be addressed. However, the blunters, who may prefer getting some information but not all, or none at all, should have their choice respected.

5.3.2 IMPLICATIONS FOR INFORMATION PROVISION TO HEALTH WORKERS

Although some library services had been superseded by ICTs in the developed world, they were still needed in the study areas. Specifically, health workers expressed a need for a functional and modern medical/health library in each district to ease access to information. Such a modern information unit should be accessible electronically and be prepared to proactively meet the growing demand for information services from users with ICT infrastructural challenges. The library should also identify the key reference sources for clinical problem-solving that are not indexed in international bibliographic databases. Access tools to support and ease the retrieval of such un-indexed materials should be put in place.

In addition, a list of documents that health workers mostly used or referred to in their work was identified in Chapter Two (Section 2.4.1, Information Sources). That list could be subjected to a quantitative survey to be able to generalize the findings. The final list could then be compared with that of the Blue Trunk Library collection (supported by WHO). If some titles were missing, then a proposal should be made to WHO to include them because they would be some of the most commonly used documents by rural health workers. Furthermore, given the increasing use of mobile devices by health workers, some of the basic documents should be made available electronically to ease regular updating.

Health information provided via mass media, particularly radio and television programs presented by medical doctors, was considered informative and valuable by lower-level health workers who were sometimes too busy to listen or watch. Such programs could be replayed or repackaged to increase access and use.

Training and information skills for health workers

The need for training schools to inculcate a culture of life-long learning into the trainee health workers to enhance their information seeking after training was highlighted by all the doctors interviewed. Furthermore, the medical and paramedical (nurses, midwives, clinical officers) schools' curricula should include training in informatics to improve, among other things, the IT literacy of their graduates.

For the health workers who are already in the field, there was a need to set-up postqualification programs to provide basic information skills to them. Regular outreach health information literacy sessions, such as those conducted by Makerere University librarians and health professionals in Uganda, were commendable.

Besides the information skills, health workers reported some inadequacies in their initial training that should be addressed by the medical and paramedical schools. For example, doctors reported that the undergraduate course in psychiatry did not prepare them well enough to handle the various cases they came across in practice. The clinical officers made similar comments about nutrition, and the nurses and midwives raised similar concerns about diabetes and heart diseases.

5.3.3 IMPLICATIONS FOR HEALTH INFORMATION PROFESSIONALS

It was observed that the lack of trained librarians, which used to be a major concern before the 1990s, was generally no longer a problem in Sub-Saharan Africa (Rosenberg, 1998). Although this was generally true, there was still a shortage of qualified medical/health information professionals/librarians in Sub-Saharan Africa. This work confirmed that the challenges facing medical/health information units in Uganda were compounded by a shortage of qualified medical/health information professionals. Consequently, the author introduced an elective course titled "Health information systems and services" at the graduate level as part of her postdoctoral activities. The course was conducted at Makerere University, School of Library and Information Science (LIS), and more than twenty graduate librarians had attended it. The course graduates were trained and prepared to play a key role in the provision of health information and the management of health information systems. There was demand for this or similar courses to be introduced at undergraduate or graduate levels in other LIS institutions in Sub-Saharan Africa, as recommended by the Association for Health Information and Libraries in Africa (AHILA) 2008 conference. A similar recommendation was made by Kasalu and Ojiambo (2015) at the annual IFLA Health and Biosciences Libraries section conference in 2015.

Furthermore, continuing professional development (CPD) sessions are important for health information professionals to update knowledge and skills to be able to embrace the new developments/advances in librarianship, ICTs and health. Among other things, this would support the implementation of SDGs, particularly SDG number three that focuses on health, and would enhance the practice and image of librarians as professionals in a digital age.

While commenting about the Rockefeller Foundation supported health informatics training in developing countries, Hersh et al. (2010) pointed out that it was important to identify and develop the skills, training and competencies that are consistent with local cultures, languages and health systems that will be needed to realize the full benefits of ICTs. It is, therefore, important to devise training programs in Africa to address the local realities, the changing demands and the IT developments.

Some of the lessons learned by the practicing information professionals are that automation is never a completed activity because of the rapid advances in the IT environment. Librarians and other information professionals, therefore, have to keep updating their knowledge and skills. Education and CPD, which build the capacity of librarians, are important not only to the implementation of ICT-driven information activities and/or projects but also to the sustainability of the projects because the skills and knowledge acquired move the process forward and are shared with others (Musoke, 2010).

During the 2014 AHILA conference, medical librarians and other health information professionals in the WHO-AFRO region reported some achievements in addressing the changing needs of health information users. Some information professionals also demonstrated that they were prepared to modify their roles, acquire new knowledge and skills, to be able to remain relevant to the health information environment and its changing demands. It was, therefore, important to build on the achievements and share experiences and best practices through collaboration and networks with national, regional and international information organizations on health and related aspects. Projects and initiatives that had successfully addressed the information needs of health workers should be scaled up and replicated in other medical libraries/health information units. Sharing expertise with other librarians/information professionals in Africa would be cost-effective and important for the south-to-south support. This would lead to the increased access and use of quality medical/health information and, consequently, improved health outcomes.

5.4 **AREAS FOR FURTHER RESEARCH**

Theory is a process, an ever-developing entity, and not a perfected product (Glaser and Strauss, 1967); therefore, research has to go on. In theoretical terms, a good research product is not only one that is said to be valid but also one that is productive in terms of generating new ideas and stimulating further research (Giddens, 1997). Part of the pleasure of doing research is to see how each study leads to more and deeper questions. Research then continues as a way of answering as well as generating questions. From this book, the following issues will make the research process continue:

i. Passive access was the principle behavioral mode of health information acquisition among grassroots women leaders in rural Uganda. As already pointed out, most previous information research focused on active seeking. There was scanty literature about people who do not engage in active information seeking, hence the need for further research on this topic.

   Furthermore, some women interviewees pointed out that they were socialized to be generally passive. Research to find out whether grassroots male leaders in rural areas would exhibit a different type of health information acquisition behavior is needed.

ii. The poor reading culture of lower-level health workers reported in Chapter Two is also an area for further investigation. This would establish, for example, whether the poor reading habits of nurses, midwives and other lower-level health workers are due to the following: educational level; lack of information skills in their training/curriculum; work load in health units; nursing or midwifery being a practical discipline, once one masters the skill that would be good enough; limited access to information materials and a combination of these factors or something else.

iii. In view of the fact that most previous studies about information seeking behavior had focused on academics or urban-based professionals, there is a need for research on the information behavior of rural-based professionals in Africa.

iv. The Interaction-Value model, that emerged from a qualitative study, has been able to illuminate issues beyond those that the study was originally designed to understand. For example, in addition to gaining insights into the information processes and behaviors surrounding access to and use of health information, the Interaction-Value model highlights the meaning and value of information derived from the interactions between women with other women, LCs, health workers and church leaders, and between the health workers with their juniors, seniors and members of professional associations, which give it a symbolic interactionist perspective. The potential use of this social psychological theoretical approach to information behavior is something that needs further attention from information researchers.

v. The findings in this book shed some light on the information needs and information behavior of people with long-term illnesses, such as sickle cell anemia, asthma, AIDS and paralysis. These findings emerged from the interviews. There is, therefore, a need for further and more specific research on coping and information behavior in Uganda. (Baker (1995, 2005) also observed that because few studies had focused on people with chronic diseases, information about monitoring and blunting behavior in the face of long-term stress remained relatively scarce. It was, therefore, recommended to develop a body of knowledge about the information needs and information seeking behaviors of people with chronic diseases. LIS researchers should focus on specific diseases.

vi. The findings presented in this book could be used, in the future, to identify the changes in health outcomes and correlate them to changes in information behavior, as documented in this book. The rich data would also provide the necessary details that could later be used in large-scale quantitative surveys.

vii. The application of the Interaction-Value model is another area for further research. It was noted that hardly any research had been dedicated to health information access factors as well as to the use of information by the population investigated in the book; hence, more extensive research with larger samples should be undertaken to test the Interaction-Value model that emerged from the qualitative data. The model could also be applied to other rural areas in Sub-Saharan Africa, upper levels of PHC and/or urban Uganda, and in other parts of the world.

REFERENCES

Baker, L.M., 1995. A new method for studying patients information needs and information seeking patterns. In: Lloyd-Williams, M. (Ed.), Health Information Management Research. Proceedings of the First International Symposium. University of Sheffield, Department of Information Studies, Centre for Health Information Management Research, Sheffield, pp. 67–75, April 57.

Baker, L.M., 2005. Monitoring and blunting. In: Fisher, K., et al., (Eds.), Theories of Information Behaviour. American Society for Information Science and Technology, Medford, NJ.

Choo, C., Marton, C., 2003. Information seeking on the web by women in IT professions. Internet Res. Electron. Netw. Appl. 13 (4), 267–280.

Commission for Africa, 2007. Commission for Africa Report.

Ellis, D., 1993. Modelling the information seeking patterns of academic researchers: a Grounded theory approach. Libr. Q. 63 (4), 469–486.

Ford, N., 2015. Introduction to Information Behaviour. Facet, London.

Giddens, A., 1997. Sociological Theory. third ed. Blackwell Publishers, Oxford.

Glaser, B., Strauss, A., 1967. The Discovery of Grounded Theory: Strategies for Qualitative Research. Aldine de Gruyter, New York, NY.

Hersh, W., et al., 2010. Building a health informatics workforce in developing countries. Health Aff. 29 (2), 275–278.

IFLA, 2014. Lyon declaration on access to information and development. http://www.lyondeclaration.org/about/ (accessed 09.06.15).

Kapiriri, L., Bondy, S.J., 2006. Health practitioners' and health planners' information needs and seeking behavior for decision making in Uganda. Int. J. Med. Inform. 75 (10–11), 714–721.

Kasalu, J., Ojiambo, J., 2015. Educating health librarians in Africa today: competencies, skills and attitudes required in a changing health environment. http://library.ifla.org/id/eprint/1104.

Kinengyere, A.A., Olander, B., 2011. Users' knowledge, attitudes and practices regarding electronic resources and information literacy: a pilot study at Makerere University. In: Qualitative and Quantitative Methods in Libraries (QQML 2011) International Conference, Athens. makir.mak.ac.ug/handle/10570/1655.

Lugya, F.K., Mbawaki, I., 2011. Usability of Makula among Makerere University Library users: a case study. 3rd International Conference on Qualitative and Quantitative Methods in Libraries. National Hellenic Research Foundation, Athens, http://hdl.handle.net/10570/1585.

Mabawonku, I., 2006. The information environment of women in Nigeria's public service. J. Doc. 62 (1), 73–90.

Mbondji, P.E., et al., 2014. Resources, indicators, data management, dissemination and use in health information systems in Sub-Saharan Africa: results of a questionnaire-based survey. J. R. Soc. Med. 107, 28–33.

Musoke, Maria G.N., 2010. Reconstruction@maklib with minimal resources. http://www.ifla.org/files/hq/papers/ifla76/106-musoke-en.pdf.

Musoke, Maria G.N., 2014. Enhancing access to current literature by health workers in rural Uganda and community health problem solving. http://library.ifla.org/868/1/088-musoke-en.pdf.

Musoke, M., et al., 2014. The changing IT trends: are academic libraries coping? Qual. Quant. Methods Libr. (QQML) J. http://www.qqml.net/papers/December_2014_Issue/342QQML_Journal_2014_Musokeetal_Dec_787-809.pdf.

Namaru, R., et al., 2009. Accessibility and utilisation of HIV/AIDS information by undergraduate University students: a case study of Moi University. East Afr. J. Inf. Sci. 1 (2), ISSN: 1975–1442.

Odini, C., 1995. A Comparative Study of the Information Seeking and Communication Behaviour of the Kenya Railways and British Rail Engineers in the Work Situation. PhD. Thesis, University of Sheffield.

Omaswa, F., Crisp, N., 2014. African Health Leaders: Making Change & Claiming the Future. Oxford University Press, Oxford.

Palmer, J., 1990. Factors Affecting the Information Behaviour of Agricultural Research Scientists. PhD. Thesis, University of Sheffield.

Palsdottir, A., 2007. Patterns of information seeking behaviour: the relationship between purposive information seeking and information encountering. In: Bath, P., et al., (Eds.), Proceedings of the 12th International Symposium for Health Information Management Research (ISHMR). University of Sheffield, Sheffield, pp. 3–15.

Rockefeller Foundation, 2008. From silos to systems: Bellagio eHealth call to action on "better health for all through integrated, person-centered eHealth systems." Retrieved from: http://www.ehealth-connection.org/.

Rosenberg, D., 1998. IT and University Libraries in Africa. Electron. Netw. Appl. Policy 8 (1), 5–15.

Smith, R., 2002. Patient-oriented evidence that matters—POEM—a week. BMJ 325, 983.

Soto, S., 1992. Information in Dentistry: Patterns of Communication and Use. PhD. Thesis, University of Sheffield.

The National Planning Authority, 2013. Uganda's vision 2040. http://npa.ug/wp-content/themes/npatheme/documents/vision2040.pdf.

WHO, 2010. Connecting for Health: Global Vision, Local Insight. Report for the World Summit on the Information Society.

World Information Society, 2014. Measuring the Information Society Report.

Index

Note: Page numbers followed by "*f*" and "*t*" refer to figures and tables, respectively.

A

Academic work, 124
Access to Global online Research in Agriculture
  (AGORA), 10
Access to information, 1, 12–13
  constraints to, 138
  for health workers, 67, 67*t*
  for women, 24–25
Access to Research for Development and Innovation
  (ARDI), 10
Actions of health workers, 122–126
  detection, 126
  information dissemination, 123–125
    academic work, 124
    preventive work, 123–124
    professional support, 124–125
    making decisions, 125–126
Active information needs, 35–36
Active information seeking, 41, 77–78, 139
  use of radio, TV and newspapers in, 28
Actual sources of information
  for health workers, 59–90, 60*t*
    books and periodicals, 60–61
    community-based agencies, 66–67
    formal and online information sources, 62–65
    media, 65–66
    most important information sources, 59–60, 60*t*
    professionals (seniors and colleagues) and
      associations, 62
    seminars, workshops and other training sessions, 61
  for women, 22–24, 23*t*
AGORA. *See* Access to Global online Research in Agriculture
  (AGORA)
AHILA. *See* Association for Health Information and Libraries
  (AHILA)
Albert Cook Library, 10
Alcoholism, 116
Alma-Ata Declaration, 6
Antimalarial drugs, 108–109
Apathy, 48–49, 82
"Appropriate prescription/medicine," information needs for,
  33–34
ARDI. *See* Access to Research for Development and
  Innovation (ARDI)
Association for Health Information and Libraries
  (AHILA), 162
Attitudinal change, 110
Awareness raising, 112–115

B

Becker Medical Library model, 144
Behavioral change, 115–117

C

Capacity building information needs, 38
Categories, 21–22
  constraints, 22
  information needs, 21
  information sources, 21
  moderators, 22
  value of information, 22
Choice of information source, 25–27
Church leaders, 54
Clinical information needs, 73–74
Clinical work of health workers, 121–122
COBES. *See* Community-based education and service
  program (COBES)
Collaboration/cross-fertilization, 54
Collection of dissimilar attributes, 132
College of Health Sciences at Makerere University
  (Mak-CHS), 10
Community-based agencies, 66–67
Community-based education and service program
  (COBES), 6
Community mobilization and sensitization, 130–131
Community support, 34
  and self-help, 118–119
Connotations of value, 67
Consortium of Uganda University Libraries (CUUL), 10
Constant comparative method, 20
Constraints, 42–51, 96–100
  health workers, information activities, 78–84
    enhancing/supportive factors, 79
    to information access, 79–82
    to information use, 82–84
    negating/hindering factors, 79
  overcoming, 111
  women, information activities by, 42–51
    analytical representation of, 44*f*
    to information use, 49–51
    negating factors, 42–43
    supportive factors, 43–49
Constraints to information access, 79–82
  concerning specific information sources, 79–81
  moderators of, 52–58, 85–89
  socioeconomic, 45–49, 81–82

Constraints to information access and use by women, 43, 44*f*
  constraints concerning specific information sources, 43—45
  socioeconomic constraints to information access, 45—49
Constraints to information use
  for health workers, 82—84
    changes in medical practice, 83—84
    medicine/facilities, unavailability of, 84
    quality of information accessed, 82—83
  for women, 49—51
    sociocultural constraints, 50—51
Continuing medical education (CME) materials, 74, 121—122
Continuing professional development (CPD), 72, 74—75, 78,
    84, 90, 130, 162
Contraceptives, use of, 38
Coping, 35, 111
  and information behavior, 164
  with stress, 35
CPD. *See* Continuing professional development (CPD)
Credibility/reliability, 25—26
Critical case sampling strategy, 18
Critical incident technique, 19—20
Critical information needs, 32—35, 40
Cultural taboos, 151—152

D
Data collection process, 19
Decentralization, 5
Dervin's theory, 148—149
Developing countries, 1
  Internet connectivity in, 9
  PHC strategy in, 4
Diagrams, 137
Disseminated information, 146
Document delivery service (DDS), 70

E
Ebola, 95, 129
Economic moderators, 58, 87—89
Economic-related constraints, 46
Educational constraint, 47—48
Educational moderators, 55—56, 89
Education and training information needs, 74—76
eHealth, 159
Elsevier Foundation, 127
Emotion-focused coping, 35—36

F
Face-to-face interaction, 145—146, 157
Face-to-face interviews, 19, 78
Family planning (FP), 38, 50, 110, 117
Financial constraints, 81
FM stations, 55

Formal and online information sources, 62—65
Formal methods of information delivery, 112—114

G
Gender and culture, 47
Grounded model, 143
Grounded theory, 17—18, 20, 142—143, 151

H
Health care, effect of information on, 126—131
  background, 126—127
  evidence of, 127—131
    examples, by health workers, 127—129
    examples, by women, 129
    proposed ways of measuring, 130—131
Health decisions, 117—118
Health education, 126, 161
Health informatics, implications for, 158—163
Health Informatics Building Blocks (HIBBs), 159
Health information, 1
  challenges in, 11—13
  information concept, 2—4
  information technology, in health sector and related
      policies, 8—11
  primary health care, 4—7
Health information challenges, 11—13
Health Information Management System (HMIS), 8, 68, 77
Health information professionals, implications for, 162—163
Health InterNetwork Access to Research Initiative (HINARI),
    10, 130, 159
Health knowledge, 38
  and improved health, 109—110
Health Planning Unit, 8
Health workers, 18, 139—140, 144, 148—149
  information activities by. *See* Information activities: by
      health workers
  value of information category and subdivisions, 120*f*
HIBBs. *See* Health Informatics Building Blocks (HIBBs)
HINARI. *See* Health InterNetwork Access to Research
    Initiative (HINARI)
HIV/AIDS, 31, 39, 55, 65, 73, 116
  awareness about, 116—117
  information about, 129
  testing, 129
HMIS. *See* Health Information Management System (HMIS)
Holistic inductive paradigm, 13, 17
Hospital libraries, 62—64
Human information behavior, 143

I
ICTs. *See* Information and communication technologies
    (ICTs)

IFLA. *See* International Federation of Library Associations (IFLA)

Illiteracy, 92

Illness, type of, 26–27, 33
childhood, 37

Immediate feedback, 91

Immunization, 36, 40, 51, 99–100, 107, 110, 117–118, 129

Implications for theory, practice and the future, 155
and applicability of findings, 156–158
areas for further research, 163–164
health informatics and information provision in general, 158–163
health information professionals, 162–163
information provision to health workers, 161–162
information provision to women, 160–161

In-between approaches, 143

Infant mortality and morbidity, reduction of, 130–131

Informal communication networks, 115

Informal health information, 115, 123

Informal information exchange routes, 115

Informal methods of information delivery, 112, 114–115

Information, education and communication (IEC) activities, 127, 135

Information, value and impact of, 131–134

Information acquisition, 28, 40–41, 77

Information activities
by health workers, 59–90
analytical interpretation of, 68*f*
constraints, 78–84
information needs, 70–78
information sources, 59–70
moderators, 84–90
by women, 22–59
constraints, 42–51
information needs, 28–42
information sources, 22–28
moderators, 51–59

Information and communication technologies (ICTs), 8, 159

Information behavior, 3–4, 13, 105–106
of health workers, 70, 77–78
active information seeking, 77–78
of women, 27–28, 38–42
active and passive, 39–41
passive access, 42
passive versus active access, 28
related to coping with health problems, 39
semi-active seeking behavior, 41
use of radio, TV and newspapers, 28, 65

Information behavior, modeling, 137
core and main categories, 137–139
interaction value model, 139–142
discussion of, 142–150
limitations of, 151–152
stage four, 141*f*, 142

stage one, 140*f*, 141–142
stage three, 141*f*, 142
stage two, 140*f*, 142
strengths of, 150

Information concept, 2–4

Information dissemination, 105–106, 112–115, 123–125, 138, 147

Information encountering, 40–41

Information gatekeepers, 18

Information measure, 3

Information needs, 21, 94–95, 137–138
of health workers, 70–78
analytical subdivisions of, 72–77
information behavior, 77–78
information that had been accessed, 71
information that was not fully accessed, 71
of women, 28–42
analytical insights and interpretation of the data on, 31–38
information behavior, 38–42
information that was accessed, 30–31
information that was not accessed, 29–30

Information provision in general, implications for, 158–163

Information provision to health workers, implications for, 161–162

Information provision to women, implications for, 160–161

Information-related behavior, 17

Information-related challenges in the health sector, 11

Information-seeking behavior, 11, 13, 39–41, 77, 87, 143, 148–149, 156

Information sources, 21, 90–94, 137–138
for health workers, 59–70
actual sources, 59–67
analytical subdivisions of, 68–70
best/easiest ways to access health information, 67
potential sources, 67
for women, 22–28, 23*t*
actual sources, 22–24
analytical subdivisions, 25–27
best/easiest ways to access health information, 24–25
information behavior, 27–28
potential sources, 24

Information technology, in health sector and related policies, 8–11

Information use, constraints to, 138

Information use and attribution of value by health workers, 119–126
actions, 122–126
detection, 126
information dissemination, 123–125
making decisions, 125–126
value of information, 119–122
administration, 122
clinical work, 121–122

Information use and attribution of value by
women, 105–119
actions, 111–119
community support and self-help, 118–119
information dissemination and awareness raising,
participation in, 112–115
making decisions, 115–118
value of information, 106–111
causes and prevention, 107–108
constraints, overcoming, 111
coping, 111
health knowledge and improved health, 109–110
misconceptions, overcoming, 110–111
treatment, 108–109
Innovative Libraries in Developing Countries (ILDC)
program, 127
In-service training sessions, 86
Integrated Management of Childhood Illnesses
(IMCI), 121
Interactionist approaches, 143
Interaction value model, 141*f*, 142–143, 149–150, 158, 164
development of, 139–142
stage four, 141*f*, 142
stage one, 140*f*, 141–142
stage three, 141*f*, 142
stage two, 140*f*, 142
discussion of, 142–150
limitations of, 151–152
strengths of, 150
Interactivity, issue of, 91
International Federation of Library Associations (IFLA), 10,
127, 132, 160
Internet, 9, 24, 45, 64–65, 93, 98, 115
Interpersonal interactions, 145–146
Interpersonal moderators, 54–55, 85–86
Interpersonal sources, 26
Interview guide method, 18–19

L

Language differences, 151
Latent information needs, 36–38, 40
Latent needs, 32, 95
Leaders as moderators, 52–54
Libraries, 11, 24, 45, 62–63, 92–93, 161
hospital libraries, 63
Literacy, 55
and access to information, 56
and formal education, 55
Literacy and formal education, 55–56
Local councils (LCs), 23, 28, 43, 52–54, 66–67, 101,
112–113
Local initiatives, 89–90, 101
Lyon Declaration (2014), 160

M

MADDO (Masaka Diocesan Development Office), 123–124
Makerere University Library (Maklib), 10, 56, 70, 89–91, 93,
162
Making decisions, 115–118, 125–126
behavioral change, 115–117
treatment choices, 117
Malaria, 29, 107–109
Male research assistants, 19, 151–152
Mathematical theory of communication, 3
Mbarara University of Science and Technology (MUST), 6,
93
Measles, 29, 128
Media, 65–66
Medical doctors, 21, 30, 60–61, 78, 89–90, 93
Medicine/facilities, unavailability of, 84
Memos, 137
Methodological approach to research process, 17–20
Met needs, 32
Ministry of Health (MoH), 5, 8, 57–58, 73–74, 97, 129,
160–161
Misconceptions, overcoming, 38, 110–111
Mobile phone, 24, 57, 61, 67, 93
Mobile technologies, 160
Moderation of constraints, 101, 139, 142
Moderators, 22, 51–59, 84–90, 85*f*, 100–102, 138
of constraints to information access, 52–58, 85–89
of constraints to information use, 58–59, 89–90
for health workers, 89–90
for women, 58–59
economic, 58, 87–89
educational, 55–56, 89
interpersonal, 54–55, 85–86
leaders as, 52–54
personal, 54–55, 85
professional, 86–87
rural outreach moderators, 55
social, 57
spatial/geographical, 56
technological, 57–58
Modernization theories, 33
MoH. *See* Ministry of Health (MoH)
Monitoring and blunting theory, 39, 149
MTrack, 57–58, 65
Multiple sclerosis, 39
Multisectoral approach, 6–7
MUST. *See* Mbarara University of Science and Technology
(MUST)

N

National Research and Education networks (NRENs), 94
Needs for health information, 28–29, 30*t*, 70
Negating factors, 42–43, 79

Newspaper clippings on health, posting, 126
Normative value approach, 131–132
Nursing assistants, 18

O

OARE. *See* Online Access to Research in the Environment
        (OARE)
Objective information, 3–4
Online Access to Research in the Environment (OARE), 10
Online information sources, 64–65
Over-the-counter drugs/treatment (OCTs), 26

P

Passive access to information, 42, 78, 148, 163
Passive attention, 28
Passivity, 22, 82
PEAP. *See* Poverty Eradication Action Plan (PEAP)
Perceived value approach, 131–132
Personal actions, 122–123
Personal health, 125
Personal moderators, 54–55, 85–86
PHC. *See* Primary health care (PHC)
Places of worship, 26
Planned health strategies, 11
Policy decisions, need for, 7
Polio immunization, 51, 101, 107, 117–118
Potential sources, of information
    for health workers, 67
    for women, 24
Poverty Eradication Action Plan (PEAP), 11
Preventive care information needs, 76
Preventive work, 123–124
Primary health care (PHC), 2, 4–7, 18, 34, 143, 155
    principles, 5–6
    services, 6f
Primary noninformation function, 23
Printed health information, 46
Printed information, 30, 67, 92–93
Print sources, 61, 91–92
Problem-focused coping, 35–36
Professional associations/NGOs, 62
Professional moderators, 86–87
Professional support, 86, 124–125
Profession-related constraints, 81–82
Psychological and physical suffering, 111

Q

Qualitative analysis, 17–18
Qualitative research, 19–20, 158
Quality of information, 29, 58–59, 107, 110

R

Radio, 22–24, 26, 45, 55, 65–66, 89, 91
Reciprocity, 26
    lack of, 25
Religious leaders, 53–54, 147
Religious practices, 47, 50–51
Research fatigue, 152
Research for Life (R4L), 10
Rockefeller Foundation, 159, 162
Rural health workers, 61, 79
    actual information sources for, 60t
    best and easiest ways to provide information to, 67t
    enhancing/supportive factors, 79, 79t
    negating/hindering factors, 79, 79t
Rural Nigerian setting, 23
Rural outreach moderators, 55
Rural women's access to health information, factors
        affecting, 42t

S

Scares/warning bells, 108
SDGs. *See* Sustainable Development Goals (SDGs)
Self-concept/perception, 49
Self-control, 50–51
Self-help, 118–119
Seminars/workshops, 28, 46–47, 56, 61, 91
Sense-making theory, 148–149
Sensitivity test, 23
Sensitization and marketing of services/facilities, 123–124
Sexually transmitted infections (STI), 50
Shannon's model, 4
Sickle cell anemia, 35–36
Sickle cell disease attacks, 111
Slawson and Shaugnessy formula, 158
SNA. *See* Social network analysis (SNA)
Social constraints, 46, 50–51, 82
Social moderators, 53, 57
Social network analysis (SNA), 147–148
Social networks, 100, 115, 147
Social practice and activities, 145–146
Socioeconomic constraints to information access, 45–49,
        81–82
Spatial factors, 46
Spatial/geographical moderators, 56
Steps to Behavior Change (SBC) model, 110–111
"Straight talk", 114–115
Subjective information, 3–4
Sub-Saharan Africa, 1–2, 7, 10–12
    health information challenges, 11, 13
    Internet connectivity in, 9
    IT-related studies on health, 10–11
Supportive factors, 43–49, 79
Supportive infrastructure, 59

Sustainable Development Goals (SDGs), 160, 162
Symbolic interactionism, 145–146

T

TBAs. *See* Traditional birth attendants (TBAs)
Technological moderators, 57–58, 88–89
"Technology, People and Process" (TPP) program, 9
Television, 28, 89
Theoretical sampling strategy, 17–18
Thick description, 22, 152
Time management, lack of, 47
Traditional birth attendants (TBAs), 18–19, 77–78, 157
Training and information needs, 75
Training and information skills, for health workers, 162
Training programs for rural people, 56
Transport facilitation, 87–88
Treatment choices, 117
Treatment information, 37–38, 108–109

U

UCH. *See* Uganda Chartered Healthnet (UCH)
Uganda Chartered Healthnet (UCH), 10, 160
Uganda Communications Act, 9
Uganda Communications Commission, 9
Uganda Health Information Digest, 60–61, 63–64, 90
Uganda Health Information Network (UHIN), 10, 90
Uganda national AIDS awareness program, 117
UHIN. *See* Uganda Health Information Network (UHIN)
Unhealthy lifestyles, globalization of, 7
Unmet health information needs, 32, 95, 99–100
Unmet needs, 30, 32, 95, 138

V

Value and impact of information, 131–134
Value of information, 22, 139
   and actions, 111–119, 138–139
   core category, 105
   health workers, 119–122
      administration, 122
      clinical work, 121–122
   and resultant actions, 115*f*
   women, 106–111
      causes and prevention, 107–108
      constraints, overcoming, 111
      coping, 111
      health knowledge and improved health, 109–110
      misconceptions, overcoming, 110–111
      treatment, 108–109
Village Health Team, 4, 38
"Vision 2040", 160

W

WHO. *See* World Health Organization (WHO)
Women, information activities by. *See* Information activities,
      by women
Women's actions, analytical interpretation of, 112*f*
Women's information behavior, 40, 115, 139, 148–149
Women's participation in health-related events, 55
Women's value of information category, 106*f*
World Health Organization (WHO), 7, 97, 133

Printed in the United States
By Bookmasters